Computer Sound Design

CD

7
Day
Loan

Music TECHNOLOGY Series

Titles in the Series

Computer Sound Design

Synthesis techniques and programming

Second edition

Eduardo Reck Miranda

AMSTERDAM • BOSTON • HEIDELBERG • LONDON • NEW YORK • OXFORD
PARIS • SAN DIEGO • SAN FRANCISCO • SINGAPORE • SYDNEY • TOKYO
Focal Press is an imprint of Elsevier

Focal Press is an imprint of Elsevier
Linacre House, Jordan Hill, Oxford OX2 8DP, UK
30 Corporate Drive, Suite 400, Burlington, MA 01803, USA

First edition published as *Computer Sound Synthesis for the Electronic Musician* 1998
Second edition 2002
Reprinted 2004, 2005, 2006

Notice
No responsibility is assumed by the publisher for any injury and/or damage to persons
or property as a matter of products liability, negligence or otherwise, or from any use
or operation of any methods, products, instructions or ideas contained in the material
herein. Because of rapid advances in the medical sciences, in particular, independent
verification of diagnoses and drug dosages should be made

British Library Cataloguing in Publication Data
Miranda, Eduardo Reck, 1963–
   Computer sound design: synthesis techniques and
   programming. – 2nd ed. – (Music technology series)
   1. Computer sound processing   2. Music – Data processing
   I. Title   II. Computer sound synthesis for the electronic musician
   786.7′6

Library of Congress Cataloging-in-Publication Data
A catalog record for this book is available from the Library of Congress

ISBN–13: 978-0-240-51693-6
ISBN–10: 0-240-51693-1

For information on all Focal Press publications
visit our website at www.focalpress.com

Printed and bound in *Great Britain*

06 07 08 09 10   10 9 8 7 6 5 4

Contents

Series introduction

The Focal Press Music Technology Series is intended to fill a growing need for authoritative books to support college and university courses in music technology, sound recording, multimedia and their related fields. The books will also be of value to professionals already working in these areas and who want either to update their knowledge or to familiarise themselves with topics that have not been part of their mainstream occupations.

Information technology and digital systems are now widely used in the production of sound and in the composition of music for a wide range of end uses. Those working in these fields need to understand the principles of sound, musical acoustics, sound synthesis, digital audio, video and computer systems. This is a tall order, but people with this breadth of knowledge are increasingly sought after by employers. The series will explain the technology and techniques in a manner which is both readable and factually concise, avoiding the chattiness, informality and technical woolliness of many books on music technology. The authors are all experts in their fields and many come from teaching and research backgrounds.

Dr Francis Rumsey
Series Consultant

Foreword

By Jean-Claude Risset

The book *Computer Sound Design: Synthesis techniques and programming* by Eduardo Reck Miranda is most welcome – it should be very helpful to anyone concerned with music technology.

During the last century, electricity has gradually permeated the world of sound. Most music is now heard through loudspeakers. But, beyond reproduction, electrical technology has also brought radically new resources for the production of new sounds and music.

It took time for the musical world to become aware of the potential of these resources. Yet, for many years, Edgard Varèse tried to persuade researchers and institutions to take advantage of science and technology for developing sound producing machines – only at the end of his life was he able to compose tape music and encourage the first steps of computer music. Iannis Xenakis called upon the resources of the computer for several of his compositions and Pierre Boulez initiated IRCAM, an institute dedicated to musical research and creation and which mostly uses digital tools. Popular music could never have evolved as it has done without electronic equipment and the quest of groups for developing their own 'sound' involves research into sound design. Thus the new possibilities have already had a far-reaching influence on music composition. The exploration of these novel possibilities is still going on: it is the most exciting challenge for musicians today.

Advanced electronic technology for sound has long been 'analogue'; it has opened new possibilities – 'musique concrète', made up from recorded acoustic sounds, and 'electronic music', resorting to sounds produced electronically. However, mastering analogue technology demanded artful wizardry, so musicians generally used ready-made possibilities rather than developing their own. Moreover the equipment tended to be unstable and drifting, and it rapidly became obsolete.

Since the late 1950s, digital technology of sound, introduced by Max Mathews, has gradually replaced analogue technology. Digital coding brings precision, stability and reproducibility to sound synthesis and processing, and it allows sound control to benefit from the boundless possibilities of the general-purpose computer. As François Bayle often says, the computer is not a tool, it is rather akin to a workshop – everyone can mould it to define one's own world, to design one's own timbres, to build one's own tools, both intellectual and material, shaped according to the individual's inclinations. What is at stake here is very important – the digital domain permits the choice of specific constraints and the extension of compositional control to the level of the sound itself, not merely composing with sounds, but composing the sounds themselves.

With digital technology, a promising and unbound world of new sonic and musical possibilities appears within reach. However, this world must be explored. I myself have been so attracted to it that I have spent a great deal of my time in this exploration. John Chowning had the same inclination and he not only developed a powerful new method for sound synthesis (i.e. FM), but also created new and convincing musical works, and fostered our understanding of musical sound and its perception.

However exciting, this phase of exploration was demanding, and many musicians were unwilling to devote much time to exploring and developing new possibilities. Although powerful, synthesis software was somewhat hard to use on large computers, which were not very user-friendly. When digital real-time sound systems appeared about 25 years ago, many musicians thought they could use real-time controls, their ear and intuition to choose empirically among the possibilities offered for sound. This was largely a delusion, however. Twisting knobs, fiddling with controls and groping blindly rarely bring you what you want – just as the manipulation of a Rubik's cube without strategies could last forever without ever reaching the solution. With early real-time systems, musicians could only use in real-time the possibilities set ahead of real-time, often by designers who did not have specific aesthetic demands; in particular they tended to use ready-made sounds and factory timbres, rather than working on the sound quality. One could witness an aesthetic regression – most musicians could not resist the temptation of real-time and resorted to similar immediate possibilities, which soon became stereotypes and clichés. Moreover, real-time operation needed special-purpose processors using the most advanced microelectronic technology, but this technology evolved so rapidly that the lifetime of these processors was only a few years, and the musical works had to be adapted to newer systems. Such adaptations required a long, costly and uninspiring toil, which was often prohibitive; many musical works realized then can no longer be played today because of this problem of obsolescence.

Fortunately the working conditions for computer music are today much more favourable. Personal computers have become so powerful that they no longer require special-purpose circuits for real-time operation; the issue only concerns software – it is much easier to port from one computer to another one or to a future model. Unlike the 1960s and 1970s, when only big institutions could provide the latest resources for electroacoustic music, the most advanced tools for programming synthesis and the processing of sound can now be implemented on individual computers. Musicians can – and should – get involved in selecting and adapting these tools to their specific tastes and purposes. One can take advantage of the know-how developed by the computer music community, a know-how which can easily be transmitted in digital form – for instance in terms of 'recipes', i.e. data required for the synthesis of certain sounds that are rather difficult to obtain – this is a great asset.

Thus this book on computer sound design is timely indeed. It provides the necessary background and explanations for anyone wishing to develop or at least understand methods for designing and programming sounds. This includes sound artists, musicians – composers and performers – as well as researchers in computer music, audio and perception.

The book is designed to be very practical. It can serve as an introduction, but also as a clarification and reminder for the already experienced reader. Eduardo Reck Miranda provides the support necessary for finding one's own path in an extremely clear fashion, so that every step appears simple. The choice of topics includes important and neatly classified methods for sound modelling. At the end, the book ventures towards 'the cutting edge' – applying artificial intelligence techniques and evolutionary computing for sound design, and taking into account the implications of parallel computing for sound synthesis; the author is himself conducting research in these promising fields, at the new frontier of computer music. The CD-ROM provides useful software as well as tutorials and examples.

Doing one's own thing with digital sound technology requires some effort indeed, but this effort is necessary to take advantage of vast potential resources in an original, personal and musical fashion, and this book should help in this worthwhile endeavour. I believe that *Computer Sound Design: Synthesis techniques and programming* will play a useful role in the diffusion of knowledge and know-how on software synthesis, and greatly facilitate the approach of digital sound design.

Jean-Claude Risset
Laboratoire de Mécanique et d'Acoustique
Centre National de la Recherche Scientifique
Marseille, France

Preface

'Computers in the future may weigh no more than 1.5 tons.'

– *Popular Mechanics*, forecasting the relentless march of science, 1949

Computer Sound Design: Synthesis techniques and programming was originally written to serve as a text or reference book for a university-level course on software synthesis or computer music, but researchers, professional musicians and enthusiasts in the field of music technology will all find this book a valuable resource.

In order to be able to make optimum use of this book, the reader should have a basic understanding of acoustics and should be familiar with some fundamental concepts of music and technology. Also, a minimum degree of computer literacy is desirable in order to take full advantage of the systems provided on the accompanying CD-ROM.

The art of sound synthesis is as important for the electronic musician as the art of orchestration is for composers of symphonic music. Both arts deal with creating sonorities for musical composition. The main difference between them is that the former may also involve the creation of the instruments themselves. Those who desire to make their own virtual orchestra of electronic instruments and produce new original sounds will find this book very useful. It examines the functioning of a variety of synthesis techniques and illustrates how to turn a personal computer into a powerful and flexible sound synthesiser. The book also discusses a number of ongoing developments that may play an important role in the future of electronic music making, including the use of artificial intelligence techniques in synthesis software and parallel computers.

The accompanying CD-ROM contains examples, complementary tutorials and a number of synthesis systems for Macintosh and PC-compatible platforms, ranging from low-level synthesis programming languages to graphic front-ends for instrument and sound design. These include fully working packages, demonstration versions of commercial software and experimental programs from major research centres in Europe and North and South America.

Computer sound design is becoming increasingly attractive to a wide range of musicians. On the one hand, and with very few exceptions, manufacturers of mainstream MIDI-based synthesisers have not yet been able to produce many known powerful synthesis techniques on an industrial scale. Moreover, the MIDI communication protocol itself is not fully capable of providing flexible and expressive parameters to control complex synthesis techniques. On the other hand, the sound processing power of the personal computer is increasing, and becoming more affordable. Computers are highly programmable and practically any personal computer can run software in real time, capable of synthesising sounds using any technique that one could imagine.

Musicians often may not wish to use preset timbres but would prefer to create their own instruments. There are, however, a number of ways to program instruments on a computer and the choice of a suitable synthesis technique is crucial for effective results. Some techniques may perform better than others for some specific timbres, but there are no definite criteria for selection; it is basically a matter of experience and taste. In general, the more intuitive and flexible the technique, the more attractive it tends to be. For example, those techniques whose parameters provide meaningful ways to design instruments and timbres are usually preferable to those whose parameters are based entirely upon abstract mathematical formulae.

A growing number of synthesis techniques have been invented and used worldwide. There is not, however, a generally agreed taxonomy to study them. Indeed, the synthesiser industry sometimes makes the situation worse by creating various marketing-oriented labels for what might essentially be the same synthesis paradigm. For the purpose of this book, the author has devised a taxonomy, assuming that synthesis techniques work based upon a *model*. For instance, some synthesis models tend to employ loose mathematical abstractions, whereas others attempt to mimic mechanical-acoustic phenomena. Synthesis techniques may thus be classified into four categories according to their modelling approach: *loose modelling*, *spectrum modelling*, *source modelling* and *time-based approaches*. It is important to observe that the boundaries of this taxonomy may overlap; some techniques may qualify for more than one class. In these cases, the author has deliberately highlighted those qualities that best fit his suggested classification.

Loose modelling techniques (discussed in Chapter 2) tend to provide synthesis parameters that bear little relation to the acoustic world. They are usually based entirely upon conceptual mathematical formulae. It is often difficult to predict the outcome and to explore the potential of a loose model. Frequency modulation (FM) is a typical example of loose modelling. FM is a powerful technique and extremely easy to program but difficult to operate; computer simulations of Yamaha's acclaimed DX7 synthesiser, for example, are relatively easy to implement on a computer. Nevertheless, it is apparent that the relationship between a timbre and its respective synthesis parameters is far from intuitive. Apart from John Chowning, the inventor of FM synthesis, only a few people have managed to master the operation of this technique.

Source modelling and spectrum modelling attempt to alleviate this problem by providing less obscure synthesis parameters; both support the incorporation of natural acoustic phenomena. The fundamental difference between source and spectrum modelling techniques is that the former tends to model a sound at its source, whilst the latter tends to model a sound at the basilar membrane of the human ear.

In general, source modelling techniques work by emulating the functioning of acoustic musical instruments (see Chapter 4). The key issue of source modelling is the emulation of acoustic sound generators rather than of the sounds themselves. For example, whilst some synthesis techniques (e.g. additive synthesis) attempt to produce a flute-like sound using methods that have little resemblance to the functioning of the flute, a source modelling technique would attempt to synthesise it by emulating a jet of air passing through a mouthpiece into a resonating pipe.

The implementation of a source model is not straightforward. However, once the model is implemented, it is not complicated to interpret the role of their synthesis parameters. Take, for example, a singing voice-like instrument. A loose model using FM, for instance, would provide relatively complex synthesis parameters, such as modulation index and frequency ratio. Conversely, a source model using waveguide filtering, for instance, would provide more easily interpreted synthesis parameters, such as air pressure, vocal tract shape and throat radiation output.

Spectrum modelling techniques have their origins in Fourier's Theorem and the additive synthesis technique (see Chapter 3). Fourier's Theorem states that any periodic waveform can be modelled as a sum of partials at various amplitude envelopes and time-varying frequencies. Additive synthesis is accepted as perhaps the most powerful and flexible spectrum modelling method, but it is difficult and expensive to run. Musical timbres are composed of dozens of time-varying partials, including harmonic, non-harmonic and noise components. It would require dozens of oscillators, noise generators and envelopes to simulate musical timbres using the classic additive technique. The specification and control of the parameter values for these components are difficult and time-consuming. Alternative methods have been proposed to improve this situation, by providing tools to produce the synthesis parameters automatically from the analysis of sampled sounds. Note that the analysis techniques used here store filter coefficients rather than samples. The great advantage of spectral modelling over plain sampling is that musicians can manipulate these coefficients in a variety of ways in order to create new sounds. Sound morphing, for example, can be achieved by varying the coefficients accordingly.

Finally, time-based techniques approach synthesis from a time domain perspective. The parameters of time-based synthesis tend to describe sound evolution and transformation of time-related features; e.g. in terms of time lapses. Examples of time modelling techniques include granular synthesis and sequential waveform composition; time modelling techniques are discussed in Chapter 5.

I would like to express my gratitude to all contributors who have kindly provided the materials for the CD-ROM: Richard Moore (University of California, San Diego), Robert Thompson (Georgia State University), Kenny McAlpine (University of Glasgow), Stephan Schmitt and Jake Mandell (Native Instruments), Nicolas Fournel (Synoptic), Karnataka Group, Dave Smith (Seer Systems), Aluizio Arcela (University of Brasília), Xavier Rodet and Adrien Lefevre (Ircam), Trevor Wishart, Archer Endrich, Richard Dobson and Robert Fraser (CDP), Jacques Chareyron (University of Milan), Arun Chandra (University of Illinois),

Xavier Serra (Pompeu Fabra University), Joe Wright (Nyr Sound), Joerg Stelkens (büro </Stelkens>), Anders Jorgensen (Koblo), Claude Cadoz (Acroe), Curtis Roads (University of California, Santa Barbara), Paul Boersma (University of Amsterdam), Pedro Morales and Roger Dannenberg (Carnegie Mellon University). I would also like to thank Francis Rumsey, Robert Thompson and Dennis Miller who read the manuscript draft and made invaluable comments and suggestions for the first edition, and thanks to Peter Manning (University of Durham) for his advice on enhancing the second edition.

The front cover image was evolved using genetic algorithms technology provided by Cambrian Labs (http://www.cambrianart.com).

Many thanks go to the John Simon Guggenheim Memorial Foundation in New York for the composition Fellowship that encouraged me to embark on an in-depth study on the most wonderful of all musical instruments: *the human voice*. The synthesis of the human voice is a topic that is discussed on various occasions in this book.

Finally, I am indebted to my wife, Alexandra, for her extraordinary encouragement and support at all times. This book is dedicated to my parents, Luiz Carlos and Maria de Lurdes, who encouraged me to be a musician.

Eduardo Reck Miranda

1 Computer sound synthesis fundamentals

1.1 Digital representation of sound

Sound results from the mechanical disturbance of some object in a physical medium such as air. These mechanical disturbances generate vibrations that can be represented as electrical signals by means of devices (for example, a microphone), that convert these vibrations into time-varying voltage. The result of the conversion is called an *analog signal*. Analog signals are continuous in the sense that they consist of a continuum of values, as opposed to stepwise values.

An analog signal can be recorded onto a magnetic tape using electromagnetic technology. In order to play back sounds recorded on magnetic tape, the signal is scanned and sent to a loudspeaker that reproduces the sound vibrations in the air. Analog synthesisers basically function by creating sounds from scratch, using electronic devices capable of producing suitable signals to vibrate the loudspeakers.

Computers, however, are digital rather than analog machines. In other words, computers work based upon discrete mathematics. The adjective 'discrete' is defined in *The Collins Concise Dictionary of the English Language* as 'separate or distinct; consisting of distinct or separate parts' as opposed to continuous phenomena. Discrete mathematics is employed when entities are counted rather than weighed or measured; for example, it is suitable for tasks involving the relations between one set of objects and another set. In this case, calculations must involve only finite and exact numbers. In order to illustrate the difference between the analog and the digital domains, let us compare an analog clock with a digital clock. The analog clock does not give us a precise time value. The hands of the clock move continuously and by the time we read them, the clock has already moved forwards. Conversely, the digital clock gives a sense of precision because the digits of its display do not move continuously; they jump from one time lapse to another. Whilst the analog clock is seen to *measure* time, the digital clock is seen to *count* it.

The main difficulty in using the computer for sound synthesis is that it works only with *discrete domains* and the knowledge of sounds that science has generated throughout history is essentially analog. Moreover, computers fundamentally deal only with *binary numbers*. In contrast to the decimal numeric system, which uses ten different symbols (i.e. from 0 to 9) to represent numbers, the binary system uses only two symbols: 0 and 1. Computers are made of tiny electronic switches, each of which can be in one of two states at a time: on or off, represented by the digits '1' and '0', respectively. Consequently, the smallest unit of information that the computer can handle is the *bit*, a contraction of the term *binary digit*. For instance, the decimal numbers 0, 1, 2 and 3 are represented in the binary system as 0, 1, 10 and 11 respectively.

Computers are normally configured to function based upon strings of bits of fixed size, called *words*. For example, a computer configured for 4-bit words would represent the decimal numbers 0, 1, 2 and 3 as 0000, 0001, 0010 and 0011. Note that the maximum number that four bits can represent is 1111, which is equivalent to 15 in the decimal system. In this case, a 4-bit computer seems to be extremely limited, but in fact, even a much larger word size would present this type of limitation here. Indeed, most currently available computers use 16-, 32- or 64-bit words, but they represent numbers in a slightly different way. Computers actually consider each digit of a decimal number individually. Hence our example 4-bit computer would use two separate words to represent the decimal number 15, one for the digit 1 and another for the digit 5: 15 = 0001 0101.

The binary representation discussed above is extremely useful because computers also need to represent symbols other than numbers. Every keyboard stroke typed on a computer, be it a number, a letter of the alphabet or a command, needs to be converted into a binary number-based representation before the computer can process it. Conversely, output devices, such as the video display and the printer, need to convert the binary-coded information to suitable symbols for rendition. Manufacturers assign arbitrary codes for symbols other than numbers: for instance, the letter A = 10000001 and the letter B = 10000010. Whilst part of this codification is standard for most machines (e.g. the ASCII codification), a significant proportion is not, which leads to one of the causes of incompatibility between different makes of computer.

In order to process sounds on the computer, the analog sound signal must be converted into a digital format; that is, the sound must be represented using binary numbers. Conversely, the digital signal must be converted into analog voltage in order to play a sound from the computer. The musical computer must therefore be provided with two types of data converters: analog-to-digital (ADC) and digital-to-analog (DAC).

The conversion is based upon the concept of *sampling*. The sampling process functions by measuring the voltage (that is, the amplitude) of the continuous signal at intervals of equal duration. Each measurement value is called a *sample* and is recorded in binary format. The result of the whole sampling process is a sequence of binary numbers corresponding to the voltages at successive time lapses (Figure 1.1).

Audio samples can be stored on any digital medium, such as tape, disk or computer memory, using any recording technology available, including electromagnetic and optic technology; for example, Winchester disk drive (or hard disk), Digital Audio Tape (DAT), Compact Disc (CD), MiniDisc (MD), etc. In order to synthesise sounds from scratch, the computer must be

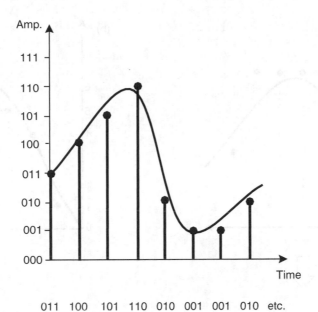

011 100   101 110 010 001 001 010 etc.

Figure 1.1 The sampling process functions by measuring the amplitude of a continuous signal at intervals of equal duration. Note that the converter rounds off the measurements in order to fit the numbers that it can process

programmed to generate the right streams of samples. The advantage of digital sound representation over analog representation is that the former allows for computer manipulation and generation of streams of samples in an infinite variety of ways. Furthermore, the possibilities of digital sound synthesis are much greater than analog synthesis.

1.1.1 The sampling theorem

The number of times the signal is sampled in each second is called the *sampling rate* (or sampling frequency) and it is measured in Hertz (Hz).

The sampling theorem states that in order to accurately represent a sound digitally, the sampling rate must be higher than at least twice the value of the highest frequency contained in the signal. The faster the sampling rate, the higher the frequency that can be represented, but the greater the demands for computer memory and power. The average upper limit of human hearing is approximately 18 kHz, which implies a minimum sampling rate of 36 kHz (i.e. 36 000 samples per second) for high fidelity. The sampling rate frequently used in computer sound design systems is 44 100 Hz.

1.1.2 Quantisation noise

The amplitude of a digital signal is represented according to the scale of a limited range of different values. This range is determined by the resolution of the ADC and DAC. The resolution of the converters depends upon the size of the word used to represent each

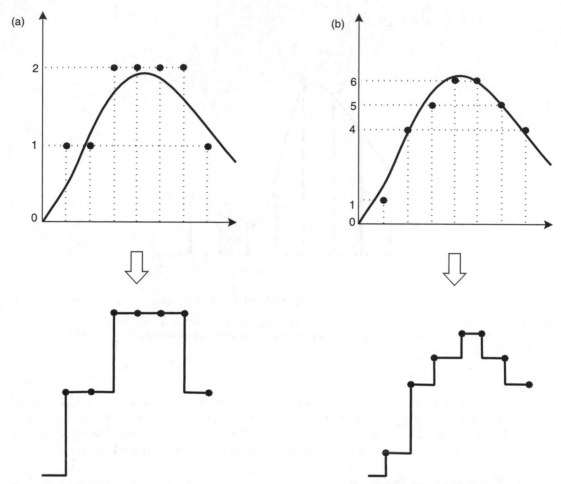

Figure 1.2 The higher the resolution of the converters, the better the accuracy of the digitised signal

sample. For instance, whereas a system with 4-bit resolution would have only 16 different values (2^4) to measure the amplitudes of the samples, a system with 16-bit resolution would have 65 536 different values (2^{16}). The higher the resolution of the converters, the better the quality of the digitised sound. The sampling process normally rounds off the measurements in order to fit the numbers that the converters can deal with (Figure 1.2). Unsatisfactory lower resolutions are prone to cause a damaging loss of sound quality, referred to as the *quantisation noise*.

1.1.3 Nyquist frequency and aliasing distortion

Nyquist frequency is the name of the highest frequency that can theoretically be represented in a digital audio system. It is calculated as half of the value of the sampling rate. For instance, if the sampling rate of the system is equal to 44 100 Hz, then the Nyquist frequency is equal to 22 050 Hz.

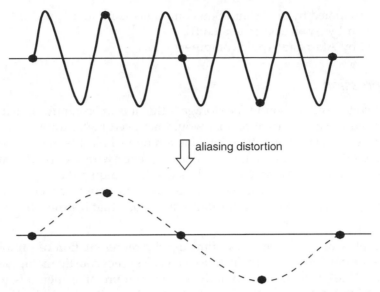

Figure 1.3 An analog-to-digital converter needs at least two samples per cycle in order to represent a sound. Otherwise the sampling process creates aliasing distortion

The ADC needs at least two samples per waveform cycle in order to represent the frequency of the sound; if not, then the frequency information is lost. Digital recording systems place a low-pass filter before the ADC in order to ensure that only signals below the Nyquist frequency enter the converter. Otherwise, the conversion process creates foldback frequencies, thus causing a phenomenon known as *aliasing distortion* (Figure 1.3).

1.1.4 Sound file formats

Sound storage

Digital audio may be stored on a computer in a variety of formats. Different systems use different sound file formats, which define how samples and other related information are organised in a computer file. The most basic way to store a sound is to take the stream of samples generated by the synthesis program, or from the ADC, and write them on a file. This method, however, is not flexible because it does not allow for the storage of information other than the raw samples; for example, the sampling rate used, the size of the word or whether the sound is mono or stereo. In order to alleviate this problem, sound files normally include a descriptive data structure, referred to as the sound file header, that encodes the properties of the file. A program that uses the sound file then reads the header and configures the system appropriately. Some sound file headers even allow for the inclusion of text comments and cue pointers in the sound. Frequently used sound file formats are:

- WAVE, adopted by Microsoft and IBM (.wav)
- VOC, adopted by Creative Lab's Sound Blaster (.voc)

- NeXT/Sun, originated by NeXT and Sun computers (.snd and .au)
- AIFF, originated by Apple computers (.aif)
- AVR, adopted by Atari and Apple computers (.avr)

Sound compression

A major disadvantage of raw sound file storage is that it is uneconomical, as it might contain a great deal of redundant information that would not need high sampling rate accuracy for representation. As a rough illustration, imagine a mono sound file, with a sampling rate of 44.1 kHz and 16-bit words, containing a sound sequence with recurrent 3-minute silences separating the sounds. In this case, 2 116 899 bits would be required to represent each of these silent portions. A compression scheme could be devised here in order to replace the raw sampling representation of the silent portions with a code that instructs the playback device to produce 3 seconds of silence.

There are a number of techniques for optimising the representation of samples in order to reduce the size of the file, some of which use clever psychoacoustic manipulations in order to trick the ear. One of the most popularly known compression methods in use today is MPEG3 (short for Moving Picture Experts Group layer 3; popularly known as MP3). Originated by the Internet Standards Organisation (ISO), MPEG3 works by eliminating sound components that would not normally be audible to humans. MPEG3 can compress a WAVE or AIFF sound file up to one-twelfth of its original size without any noticeable difference. Such compression is not, however, suitable for all kinds of music. Most discerning classic music lovers and electroacoustic music listeners do not find such compression schemes adequate.

Other compression schemes widely used today include Real Audio, created by Progressive Networks, ATRAC3, created by Sony and WMA by Microsoft. RealAudio is acknowledged as the first compression format to support live audio over the Internet. ATRAC3 is a sound compression technology based on ATRAC, the technology that Sony devised for the MiniDisc. Both ATRAC3 and WMA are identiacal to MPEG3 in many respects, with the exception that they include anti-piracy measures to inhibit illegal distribution of music over the Internet.

As far as sound synthesis is concerned, compression is normally applied to a sound after it has been synthesised. The standard practice of professional composers and music producers normally applies compression only after the piece is completed.

Sound transfer

The convenience of digital storage media is that audio samples can be transferred from one medium to another with no loss of sound quality. But the transfer of audio samples between devices can sometimes be burdensome because of eventual incompatibilities between equipment.

There are a number of data transfer formats and almost every manufacturer has established their own. From the user's point of view, it is very disappointing to learn that the new piece of equipment you always wanted is finally available for sale but that it is incompatible with the gear available in your studio. Fortunately, the digital audio equipment industry is becoming aware of these inconsistencies and is moving towards more standardised transfer formats. The two most popular transfer formats in use today are AES/EBU and S/PDIF. The AES/EBU format is a serial two-channel format, created by the Audio Engineering Society (AES) and the European Broadcast Union (EBU) for professional equipment. The S/PDIF format (Sony/Philips Digital Interface Format) is similar to the AES/EBU format, but it is intended primarily for domestic equipment and home studio set-ups. For a comprehensive coverage of this subject, refer to Rumsey and Watkinson (1995) and Rumsey (1996).

1.1.5 Super audio: Pulse Density Modulation

A trend that goes in the opposite direction to audio compression is the 1-bit audio representation, recently coined Direct Stream Digital (DSD) by Sony and Philips, also referred to as Pulse Density Modulation (PDM) by the audio engineering community. DSD uses a sampling rate of 2 822 400 MHz, that is, 64 times the standard audio CD quality of 44.1 kHz. Although this technology was originally devised for the Super Audio Compact Disc (SACD), it is likely to have a great impact on synthesis technology and computer music composition in the very near future.

Standard audio sampling uses the so called Pulse Code Modulation (PCM) technology. Starting in the late 1970s with 14-bit words, then moving up to 16-bit and 24-bit words, PCM today is capable of rather high quality digital audio coding. There is, however, increasingly little room from improvement due to a number of cumbersome technical issues. In short, PCM requires decimation filters at the sampling end and interpolation filters at the playback end of the modulation process and these add unavoidable quantisation noise to the signal. DSD eliminates the decimation and interpolation altogether, and records the pulses directly as a 1-bit signal (Figure 1.4). The analog-to-digital converter uses a negative feedback loop to accumulate the sound. If the input sound, accumulated over one sampling period, rises above the value accumulated in the negative feedback loop during previous samples, the converter outputs 1; otherwise, if the sound falls relative to the accumulated value, the converter outputs 0. As a result, full positive waveforms will all be 1s and full negative waveforms will all be 0s. The crossing point will be represented by alternating 1s and 0s. The amplitude of the original analog signal is represented by the density of pulses, hence the reason DSD is also known as Pulse Density Modulation (PDM) (Figure 1.5). The digital-to-analog conversion here can be as simple as running the pulse stream through an analog low-pass filter.

At first sight, a sampling rate of 2 822 400 Hz may seem to require prohibitive storage capacity and computer power. But this is not really the case. A standard stereo CD uses 16-bit word samples, thus the bit rate per channel here is 16 times 44.1 kHz, that is 705 600 bits per

Figure 1.4 DSD eliminates the filters and records the samples directly using a 1-bit representation. The result is a 1-bit digital representation of the audio signal. Where conventional systems decimate the 1-bit signal into multibit PCM code, DSD records the 1-bit pulses directly

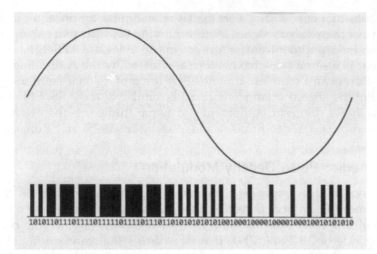

1010110111011111011111011111011101101010101010100100010000100001000100101010

Figure 1.5 In PDM, the density of pulses represents the amplitude of the sound

second. As DSD uses 1-bit per sample, the bit rate of 2 822 400 MHz is only four times higher than for a normal CD.

As far as the listener is concerned, the result of DSD gives an unparalleled impression of depth and fluidity that comes from a far greater frequency response and dynamic range. DSD captures harmonics inaudible as pure tones by accurately reproducing the leading edges of transients, which enriches the sound spectrum and enhances the musical experience.

1.2 Basics of computer programming for sound synthesis

In principle, computers can be programmed to perform or solve almost any imaginable task or problem, with the proviso that the method for resolution can be explicitly defined and the data needed for its realisation can be explicitly represented. In essence, in order to program a computer one needs to write a sequence of instructions specifying how the machine will achieve the required results. This implies that the programmer must know how to resolve the problem or task in order to instruct the machine.

In many ways, a computer program is fairly similar to a recipe for a dish: a recipe gives clear steps to prepare the dish and lists all necessary ingredients. Recipes as well as computer programs must have no ambiguities; e.g. the statement 'cover and simmer for five minutes over low heat' is preferable to 'cover and simmer for some time'. Note, however, that even though the first statement is more precise than the second, the former still implies some background knowledge to interpret it. The term 'low heat', for instance, may carry different meanings to a chef, to a chemist and to an astrophysicist; the chef will probably know better how to ascertain the gas mark of the cooker. People tend to develop specific coding and languages to communicate ideas on a specific subject more efficiently. The BBC Radio 4 Shipping Forecast is a good example of an efficient coding system to clearly communicate the weather forecast to sailors.

A programmer communicates instructions to the computer by means of a programming language. The idiosyncrasies of certain tasks require specific features which may require the design of specific programming languages to deal with them. Many programming languages have been developed for a wide range of purposes and types of computers, for example C, Basic, Fortran, Java, Lisp, Pascal and Prolog, to cite but a few. Thus, whilst Prolog was specially devised for artificial intelligence research, Java was planned for applications involving the Internet. The main advantage of using a language that has been specially designed for specific problems is that its vocabulary normally employs the jargon of the domain application. For example, the command to activate an oscillator in pcmusic (a programming language for sound synthesis; see Chapter 8) is *osc*, short for oscillator.

For the same reason as numbers and the alphabet, sounds must be converted into binary numbers before the computer can perform any processing; programs also have to be converted into machine code. Machine code is the only language that computers can understand but it is extremely tedious, if not impossible, to write a large complex program using raw binary code; hence the rationale for the so-called high-level programming languages. Such programming languages provide a 'translator' that converts the encoded program into machine code. There are two kinds of translators: *compiler* and *interpreter*. The compiler converts the whole code before executing it. Compilers save the result onto an executable file which can be activated at any time without having to repeat the translation process. Conversely, the interpreter converts each statement of the program and executes it before it proceeds to convert the next statement. Each mode has its own merits and pitfalls but an in-depth discussion of these is beyond the scope of this book.

In the celebrated book *Electronic and Computer Music*, the composer Peter Manning (1987) reminds us that the first really usable programming language for sound synthesis was Music III, created by Max Mathews and Joan Miller at the Bell Telephone Laboratories, USA, in the early 1960s (Mathews, 1969). Several subsequent versions soon appeared at Stanford University (Music 10) and Princeton University (Music 4B). Most sound synthesis languages available today are descended in one way or another from these languages, generally referred to as Music N, including the ones provided on the accompanying CD-ROM: pcmusic (developed by F. Richard Moore at the University of California in San Diego, USA), Som-A (developed by Aluizio Arcela at the University of Brasília, Brazil) and Nyquist (developed by Roger Dannemberg at the Carnegie Mellon University, USA).

1.2.1 Algorithms and control flow

An algorithm is a sequence of instructions carried out to perform a task or to solve a problem. They are normally associated with computing, but in fact they abound everywhere, even in cans of shaving foam:

```
Begin shaving foam algorithm
   Shake can well
   Wet face
   Hold can upright
   Release foam onto fingers
   Smooth onto face
```

```
   Shave
   Rinse face
End shaving foam algorithm
```

Software engineers distinguish between an algorithm and a program. An algorithm is an abstract idea, a schematic solution or method, which does not necessarily depend upon a programming language. A program is an algorithm expressed in a form suitable for execution by a computer. The program is the practical realisation of the algorithm for use with a computer.

Depending on the complexity of the task or problem, a program may require myriad tangled, complex algorithms. In order to aid the design of neat and concise algorithms, the software engineering community has developed a variety of programming schemes and abstract constructs, many of which are now embedded in most languages, including those used for sound synthesis. The more widespread of these programming schemes include *encapsulated subroutines, path selection, iteration, passing data between subroutines* and *data structures*.

Encapsulated subroutines

One of the most fundamental practices of computer programming consists of organising the program into a collection of smaller subprograms, generally referred to as *subroutines* (also known as *macro-modules, procedures* or *functions*). In this way, the same subroutine can be used more than once in a program without the need for rewriting. Subroutines may be stored in different files, frequently referred to as *libraries*; in this case, the compiler or interpreter merges the subroutines invoked by the program.

Most programming languages provide a library with a large number of subroutines which significantly facilitate the ease of the programming task. Programmers are actively encouraged to create their own library of functions as well. On highly developed programming desktops, the creation of a new program may simply require the specification of a list of subroutine calls.

In the example below, *algorithm A* is composed of a sequence of four instructions:

```
Begin algorithm A
   Instruction 1
   Instruction 2
   Instruction 3
   Instruction 4
End algorithm A
```

Suppose that this algorithm performs a task that will be requested several times by various sections of a larger program. Instead of writing the whole sequence of instructions again, the algorithm could be encapsulated into a subroutine called *Procedure A*, for example. This subroutine can now be invoked by other algorithms as many times as necessary. For example:

```
Begin algorithm B
   Instruction 5
```

```
    Procedure A
    Instruction 6
  End algorithm B
```

The method or command for encapsulating a subroutine may vary considerably from one language to another, but the idea is essentially the same in each case. As far as sound synthesis is concerned, synthesis languages provide a library of ready-made subroutines which are the building blocks used to assemble synthesis instruments.

Path selection

The instructions and subroutines must be performed in sequence and in the same order as they were specified. There are cases, however, in which an algorithm may have to select an execution path from a number of options. The most basic construct for path selection is the *if-then* construct. The following example illustrates the functioning of the *if-then* construct. Note how *Instruction 2* is executed only if 'something' is true:

```
Begin algorithm A
  Instruction 1
  IF something
    THEN Instruction 2
  Instruction 3
  Instruction 4
End algorithm A
```

Another construct for path selection is the *if-then-else* construct. This is used when the algorithm must select one of two different paths. For example:

```
Begin algorithm B
  Instruction 1
  IF something
    THEN Instruction 2
    ELSE Instruction 3
         Instruction 4
  Instruction 5
End algorithm B
```

In this case, if 'something' is true, then the algorithm executes *Instruction 2* and immediately jumps to execute *Instruction 5*. If 'something' is not true, then the algorithm skips *Instruction 2* and executes *Instruction 3* and *Instruction 4*, followed by *Instruction 5*.

The *if-then* and the *if-then-else* constructs are also useful in situations where the algorithm needs to select one of a number of alternatives. For example:

```
Begin algorithm C
  Instruction 1
  IF something
    THEN Instruction 2
  IF another thing
    THEN Instruction 3
```

```
        ELSE Instruction 4
     IF some condition
        THEN Instruction 5
     Instruction 6
  End algorithm C
```

Finally, there is the *case* construct. This construct is used when the selection of a path depends upon the multiple possibilities of one single test. In the example below, there are three 'possibilities' (X, Y and Z) that satisfy 'something'. Each 'possibility' triggers the execution of a different *Instruction*:

```
  Begin algorithm D
     Instruction 1
     CASE something
        possibility X perform Instruction 2
        possibility Y perform Instruction 3
        possibility Z perform Instruction 4
     Instruction 5
  End algorithm D
```

In this example, *algorithm D* executes either *Instruction 2*, *Instruction 3* or *Instruction 4* depending on the assessment of 'something'; only one of these three instructions will be executed.

Iteration

Iteration (also referred to as loop) allows for the repetition of a section of the program a number of times. There are many ways to set up an iteration, but all of them will invariably need the specification of either or both the amount of repetitions or a condition to terminate the iteration. The most common constructs for iteration are *do-until*, *while-do* and *for-to-do*.

The following example illustrates the *do-until* construct:

```
  Begin algorithm A
     Instruction 1
     DO Instruction 2
        Instruction 3
        Instruction 4
     UNTIL something
     Instruction 5
  End algorithm A
```

In this case, the algorithm executes *Instruction 1* and then repeats the sequence *Instruction 2*, *Instruction 3* and *Instruction 4* until 'something' is true. A variation of the above algorithm could be specified using the *while-do* construct, as follows:

```
  Begin algorithm B
     Instruction 1
     WHILE something
```

```
   DO Instruction 2
      Instruction 3
      Instruction 4
      Instruction 5
End algorithm B
```

The later algorithm (*algorithm B*) also repeats the sequence *Instruction 2*, *Instruction 3* and *Instruction 4*, but it behaves slightly different from *algorithm A*. Whereas *algorithm A* will execute the sequence of instructions at least once, indifferent to the status of 'something', *algorithm B* will execute the sequence only if 'something' is true right at the start of the iteration.

The *for-to-do* construct works similarly to the *while-do* construct. Both may present some minor differences depending on the programming language on hand. In general, the *for-to-do* construct is used when an initial state steps towards a different state. The repetition continues until the new state is eventually reached. For example:

```
Begin algorithm C
  Instruction 1
  FOR initial state TO another state
    DO Instruction 2
       Instruction 3
       Instruction 4
       Instruction 5
End algorithm C
```

In this case, the sequence *Instruction 2*, *Instruction 3* and *Instruction 4* will repeat until the state reaches 'another state'.

Passing data between subroutines

In order to construct modular programs consisting of a number of interconnected stand-alone subroutines, there must be a way to exchange information between them. As a matter of course, most subroutines (or functions) in a program need to receive data in order to be processed and should pass on the results to other subroutines. Consider this example again:

```
BEGIN algorithm B
  Instruction 5
  Procedure A
  Instruction 6
END algorithm B
```

Note that *Algorithm B* calls *Procedure A* to perform a task and they will most likely need to pass data to one another. The general passing data scheme that is commonly found in programming languages is illustrated below:

```
BEGIN fool maths
  x = 2 + 3
  y = hidden-task(x)
  z = y + 5
END fool maths
```

The *fool maths* algorithm is of the same form as *algorithm B*. After calculating a value for *x*, this value is passed on to a procedure called *hidden task*, which in turn performs a certain operation on *x*. The result of this operation is then given to *y*. Next, *y* is used to calculate the value of *z*.

The subroutine for the hidden operation must be defined elsewhere, otherwise *fool maths* will never work:

```
BEGIN hidden task(a)
   b = a × 3
END hidden task(b)
```

The role of *hidden task* is to multiply any value that is given to it. The variable inside the parentheses to the right of the name of the subroutine represents the datum being passed and it is technically called the *parameter* or *argument*. Note that the labels for the parameters do not need to match because the subroutines *fool maths* and *hidden task* are completely different from each other. However, the type (whether the parameter is a number, a word or a musical note, for example), the number of parameters, and the order in which they are specified inside the parentheses must match.

As for passing back the result, the variable on the left-hand side of the subroutine call automatically receives the result. The type of this variable must match the type of result produced by the subroutine; in this case *y* (in *fool maths*) and *b* (in *hidden task*) match their types because they are both integer numbers. The result produced by the *fool maths* algorithm is 20 (Figure 1.6).

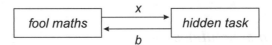

Figure 1.6 In order to construct modular programs consisting of a number of interconnected stand-alone subroutines, there must be a way to exchange information between them

Bear in mind that the actual implementation of this scheme may vary enormously, depending on the programming language used.

Data structures

In order to write algorithms that deal with data structures such as sets, matrices, lists and so on, it is often necessary to indicate clearly how these entities should be processed. The programming scheme normally used for representing and manipulating data structures in an algorithm is the *array*; almost every programming language provides ways of implementing arrays.

An array can be one-dimensional, to contain lists of things, or multidimensional, to contain matrices. An array is normally represented by a capital letter and its elements by the lower-case counterpart, furnished with an index. This index indicates the position of the element in the array. For example, an array *A* of five elements is represented as $A = [a(1), a(2), a(3),$

$a(4)$, $a(5)$]. Consider a set of composers stored in an array as follows: C = [Antunes, Arcela, Carvalho, Garcia, Iazzetta, Lintz-Maues, Richter, Silva]. In this case, the element $c(3)$ is Carvalho and Richter is $c(7)$.

In the case of a multidimensional array (e.g. to contain a matrix), the elements have one index for each dimension. For instance:

$$A = \begin{bmatrix} a(1,1) & a(1,2) \\ a(2,1) & a(2,2) \end{bmatrix}$$

Thus, the number 112 in the following array N corresponds to the element $n(1,1)$ and $n(2,2)$ = 44, to cite but two examples:

$$N = \begin{bmatrix} 112 & 54 \\ 84 & 44 \end{bmatrix}$$

In essence, arrays can be metaphorically associated with a large library bookshelf. Each book on the shelf has a code, or index. In order to retrieve a book from the shelf, librarians normally use a code that provides the coordinates to locate the book.

The following example illustrates how to fit an array into an algorithm. The algorithm below retrieves each element stored in an array of numbers and performs a multiplication on it and prints the result:

```
BEGIN multiply elements
  V[n] = [12, 6, 4, 8, 2]
  FOR x = 1 TO 5
    DO q = v[x]
       r = q × 10
       print(r)
END multiply elements
```

The array V has five elements. Each element of the array can be retrieved by an indexed variable. For example $v(1)$ = 12, $v(2)$ = 5, $v(3)$ = 4, and so on. The *for-to-do* iteration scheme is used to repeat a sequence of instructions (for multiplication and for printing) five times. At each iteration of the *for-to-do* loop, the value of x is incremental and serves as an index to retrieve the values from the array. For instance, at the third iteration x = 3 and therefore $v(3)$ = 4, which in turn is passed to the variable q. Next, r is the result from the multiplication of q by 10; this result is printed. It is assumed that the *print(r)* instruction is a subroutine defined elsewhere. When the five iterations are completed, the outcome of the algorithm will be the following values printed (either on the screen of the computer or on paper): 120, 60, 40, 80 and 20.

1.2.2 Unit generators

Synthesis languages generally provide a number of synthesis subroutines, usually referred to as *unit generators* and in this case, synthesis instruments are programmed by interconnecting a number of unit generators. Originally, the design of these unit generators was based upon the principle that they should simulate the modules of an analog synthesiser. The rationale was that electronic musicians accustomed to the setting up of analog synthesis patches would not find it so difficult to migrate to computer sound synthesis programming. Although computer synthesis technology has evolved from mere analog simulation to much more diverse and sophisticated trains of thought, a great deal of the original unit generator ideology still applies for various currently available synthesis programming languages and systems.

The most basic, but by no means the least important, unit generator of a synthesis programming language or system is the *oscillator*. The concept of a computer-based oscillator differs from the analog synthesiser's oscillator. Whereas the analog oscillator produces only a sinusoidal waveform, a computer-based oscillator can produce any waveform, including, of course, the sinusoidal one. On a computer, the oscillator works by repeating a template waveform, stored on a *lookup table* (Figure 1.7); note here the appearance of the iteration programming scheme introduced earlier. The speed at which the lookup table is scanned defines the frequency of the sound. The lookup table contains a list of samples for one cycle of a waveform, which does not necessarily need to be a sinusoid. For this reason, the specification of an oscillator on a computer always involves at least three parameters: *frequency, amplitude* and *waveform*. The most commonly used symbol for such oscillators in the sound synthesis literature is the elongated semi-circle, with the amplitude and frequency inputs at the top and the signal output at the bottom. The interior of the figure may contain a symbol or a notation defining the nature of the waveform in the lookup table (Figure 1.8).

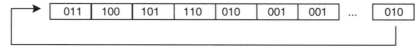

Figure 1.7 On a computer, the oscillator works by repeating a template waveform, stored in a lookup table

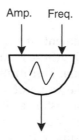

Figure 1.8 The most commonly used symbol for an oscillator

Function generators form another important class of synthesis units. These generators create the lookup tables for the oscillators and tables for sound transformation and control. Function generators fill data tables with values produced according to specific procedures or mathematical formulae, such as trigonometric functions and polynomials. The length of the function table is generally specified as a power of two, such as $2^9 = 512$ or $2^{10} = 1024$ samples.

A third class of synthesis units comprises *signal modifiers* such as low- and high-pass filters (more about filters in Chapter 4).

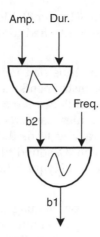

Figure 1.9 A simple instrument

The following example illustrates the codification in pcmusic of the instrument in Figure 1.9; refer to Chapter 8 for an introduction to pcmusic:

```
#include <cmusic.h>
#include <waves.h>
#instrument 0 simple;
   osc b2 p6 p5 f2 d;
   osc b1 b2 p7 f1 d;
   out b1;
end;
SINE(f1);
ENV(f2);
```

Note that the amplitude of the oscillator is controlled by an envelope, which in this case is implemented by means of another oscillator unit. The lookup table for the 'envelope oscillator' holds the shape of the envelope and is read only once per note. In this case, the frequency of the 'envelope oscillator' must be set to the inverse of the duration of the note (*freq = 1/duration*).

Envelopes are very useful for adding time-varying control to synthesis instruments and the example in Figure 1.9 illustrates only one of the many possible ways to implement them. Synthesis languages usually provide a few specially built envelope generators (e.g. *seg* in pcmusic.

1.2.3 Playing the instruments

Once an instrument has been specified, then the computer needs information to play it. The method will vary from system to system. The standard way is to write a score for the computer, as if one were to write a piano roll for a mechanical pianola. This is the case for Music N types of programming languages such as pcmusic and Som-A (both are available on the accompanying CD-ROM; refer to Chapter 8). Most synthesis systems allow for the use of controllers, such as a MIDI keyboard, attached to the computer, so that it can be played as a performance instrument.

On certain Music N type software, the musician normally creates two separate files, one containing the instrument, or a set of instruments, called an *orchestra file*, and another the parameters to 'play' the instruments, called a *score file*. The synthesis program (i.e. the compiler or the interpreter) reads both files, converts them into machine-level instructions, builds the instruments of the 'orchestra' and feeds them the parameters from the score file. The final audio signal is either written onto a sound file or is played back during processing, or both. In languages such as pcmusic and Nyquist, however, the instruments and the score are normally placed on one single file.

The main drawback of the Music N type languages is that they tend to require explicit manual specification of scores. For example, if an instrument requires 50 parameters to play a single note, then it would require the specification of 5000 parameter values to play 100 notes. The classic implementation of the popular Csound language developed at MIT Media Lab (Vercoe, 1986) can be cited as an example here. There are newer implementations of Csound that allow for real-time generation of scores on the fly using MIDI controllers (Boulanger, 2000), but the underlying score paradigm remains the same. More progressive languages alleviate this problem by providing the ability to embed formulae or functional programming facilities for generating synthesis control within the instruments themselves (e.g. Nyquist).

Those systems allowing MIDI-based control have the advantage that they can be used in conjunction with MIDI sequencing software to produce the scores. However, the limitations of the MIDI protocol itself may render this facility inappropriate for more elaborate control of synthesis instruments.

2 Loose modelling approaches: from modulation and waveshaping to Walsh and wavetable

Loose modelling techniques inherited much of their ethos from analog synthesis. At the beginning of electronic music in the 1950s, composers ventured to synthesise sounds by superimposing a few sinusoidal waves (this is known today as additive synthesis). Although very exciting to begin with, at the end of the day electronic musicians realised that in order to fulfil their increasing quest for more sophisticated sounds they would need to stockpile many more sinusoids than they had expected (refer to section 3.1, in Chapter 3). Moreover, it had become evident that in order to achieve sounds with a minimum degree of acoustic coherence, each sinusoid would need to follow its own course in terms of amplitude, frequency and phase. The technical hindrance to producing good additive synthesis at the time forced composers to look for alternative synthesis methods (Dodge and Jerse, 1985).

In this context, techniques such as amplitude modulation (AM), frequency modulation (FM) and waveshaping arose and immediately gained ground in the electronic music studio. Composers were delighted to discover that they could create rich time-varying spectra with far fewer resources than if they were to produce everything using the additive method alone. However, there was a price to pay: whereas it was straightforward to understand additive synthesis parameters such as amplitude and frequency, unfamiliar parameters such as FM's modulation index and index of deviation were not so easy to control. Indeed, as will become evident in this chapter, the only way to describe the behaviour of the resulting spectrum in these cases is by means of abstract mathematical formulations; hence the origin of the term 'loose model'.

On the whole, loose modelling techniques tend to be very easy to implement, but their synthesis parameters are not straightforward to control because they bear little relation to the acoustic world. Computer technology also inspired a few loose modelling techniques, such as Walsh synthesis and synthesis by binary instruction.

2.1 Amplitude modulation

Modulation occurs when some aspect of an audio signal (called a *carrier*) varies according to the behaviour of another audio signal (called a *modulator*). Amplitude modulation therefore occurs when a modulator drives the amplitude of a carrier. One of the pioneers of amplitude modulation in music was the composer Karlheinz Stockhausen in the 1960s (Maconie, 1976).

The *tremolo* effect may be considered to be a starting point example of amplitude modulation; it is achieved by applying a very slow sub-audio rate of amplitude variation to a sound (i.e. less than approximately 18 Hz). If the frequency of the variation is raised to the audible band (i.e. higher than approximately 18 Hz) then additional partials (or *sidebands*) will be added to the spectrum of the signal.

Simple amplitude modulation synthesis uses only two sinewave generators (or oscillators): one for the carrier and the other for the modulator. The frequency of the carrier oscillator is usually represented as f_c whilst the frequency of the modulator oscillator is represented as f_m.

Complex amplitude modulation may involve more than two signals; for example, the amplitude of *oscillator C* is modulated by the outcome of *oscillator B*, which in turn is amplitude modulated by *oscillator A*. Signals other than sinewaves (e.g. noise) may also be employed for either carriers or modulators. The more complex the signalling system, the more difficult it is to predict the outcome of the instrument. This book focuses only on simple amplitude modulation.

There are two variants of amplitude modulation: *classic amplitude modulation* (AM) and *ring modulation* (RM).

2.1.1 Classic amplitude modulation

In classic amplitude modulation (AM) the output from the modulator is added to an offset amplitude value (Figure 2.1). Note that if there is no modulation, the amplitude of the carrier would be equal to this offset value. The amplitude of the modulator is specified by an amount of the offset amplitude value in relation to a modulation index.

If the modulation index is equal to zero then there is no modulation, but if it is higher than zero then the carrier wave will take an envelope with a sinusoidal variation (Figure 2.2). In simple AM, the spectrum of the resulting signal contains energy at three frequencies: the frequency of the carrier plus two sidebands, one below and the other above the carrier's frequency value. The values of the sidebands are established by subtracting the frequency of

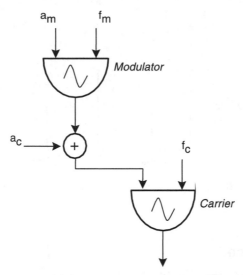

Figure 2.1 In classic AM, the output of the modulator oscillator is added to an offset amplitude-value

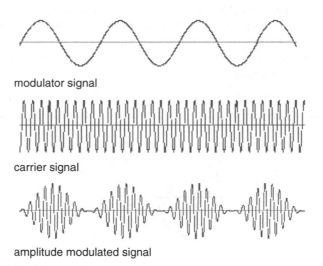

modulator signal

carrier signal

amplitude modulated signal

Figure 2.2 The amplitude of the carrier signal is controlled by the modulator signal

the modulator from the carrier and by adding the frequency of the modulator to the carrier, respectively. The amplitude of the carrier frequency remains unchanged, whilst the amplitudes of the sidebands are calculated by multiplying the amplitude of the carrier by half of the value of the modulation index, represented as *mi* (see Appendix 1 for detailed mathematical specifications). For example, when *mi* = 1, the sidebands will have 50 per cent of the amplitude of the carrier (Figure 2.3).

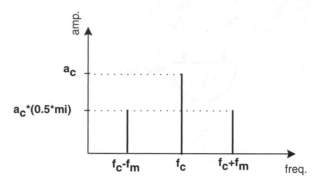

Figure 2.3 In simple AM, the spectrum of the resulting signal contains energy at the frequency of the carrier plus two sidebands

2.1.2 Ring modulation

In ring modulation (RM) the amplitude of the carrier is determined entirely by the modulator signal. Thus if there is no modulation, then there is no sound (Figure 2.4). In simple RM (i.e. when both signals are sinewaves), the resulting spectrum contains energy only at the sidebands ($f_c - f_m$ and $f_c + f_m$); the frequency of the carrier wave will not be present. RM therefore may distort the pitch of the carrier signal. For instance, if $f_c = 440\,\text{Hz}$ and $f_m = 110\,\text{Hz}$, then the instrument will produce two sidebands of $330\,\text{Hz}$ and $550\,\text{Hz}$, respectively. In RM the energy of the modulator signal is split between the two resulting sidebands (Figure 2.5). Because there is no fundamental frequency in the resulting spectrum, the sounds of RM do not usually have a strong sensation of pitch.

Ring modulation may also be achieved by the multiplication of two signals (Figure 2.6). The multiplication of two sounds results in a spectrum containing frequencies that are the sum

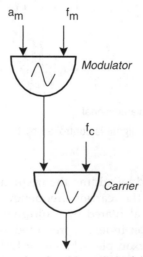

Figure 2.4 In RM the amplitude of the carrier signal is entirely determined by the modulator

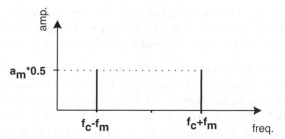

Figure 2.5 In RM, the frequency of the carrier will not be present and the amplitude of the modulator is split between the two resulting sidebands

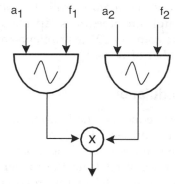

Figure 2.6 The multiplication of two signals is also a form of RM

of and difference between the frequencies of each component in the first sound and those of each component in the second.

Both AM and RM can use signals other than sinusoids, applying the same principles discussed above. In any case, great care must be taken in order to avoid aliasing distortion (i.e. generation of frequencies above 50 per cent of the sampling rate) because the highest frequencies of the two sounds will be additive.

2.1.3 Amplitude modulation examples

An example of a simple AM instrument (*Am.mae*) implemented using Audio Architect can be found on the CD-ROM (in folder *Aanets* in the Audio Architecture materials). A collection of examples in pcmusic are available in folders *Tutor* and *Ringmod*. The reader will also find a number of annotated examples on various aspects of AM and RM in Robert Thompson's pcmusic tutorial.

2.2 Frequency modulation

Synthesis by frequency modulation (FM) was originated in the late 1960s by the Stanford University composer John Chowning (Chowning, 1973; Chowning and Bristow, 1986). FM synthesis is based upon the same principles used for FM radio transmission. Chowning's first experiments employed an audio signal (called a *modulator*) to control the frequency of an oscillator (called a *carrier*) in order to produce spectrally rich sounds.

The *vibrato* effect may be considered as a starting point example to illustrate the functioning of FM. The fundamental difference, however, is that vibrato uses a sub-audio signal to modulate the carrier. A sub-audio signal is a low-frequency signal, well below the human hearing threshold which is approximately 20 Hz. The resulting sound, in this case, has a perceptibly slow variation in its pitch. If the modulator's frequency is set to a value above the human hearing threshold, then a number of partials are added to the spectrum of the carrier's sound.

There are a number of variations in FM instrument design. The most basic of the FM instruments is composed of two oscillators, called *modulator* and *carrier* oscillators, respectively (Figure 2.7). This simple architecture is capable of producing a surprisingly rich range of distinctive timbres. More complex FM instruments may employ various modulators and carriers, combined in a number of ways.

2.2.1 Simple frequency modulation

Figure 2.7 illustrates a simple frequency modulation architecture. The output of the modulator is offset by a constant, represented as f_c, and the result is then applied to control the frequency of the carrier. If the 'amplitude' of the modulator is equal to zero, then there is no modulation. In this case, the output from the carrier will be a simple sinewave at frequency f_c. Conversely, if the 'amplitude' of the modulator is greater than zero, then

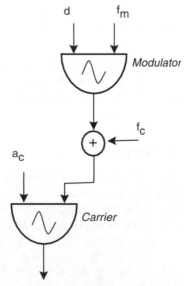

Figure 2.7 The most basic FM configuration is composed of two oscillators, called modulator and carrier oscillators. The output of the modulator is offset by a constant value and the result is applied to control the frequency of the carrier

modulation occurs and the output from the carrier will be a signal whose frequency deviates proportionally to the 'amplitude' of the modulator. The word amplitude is within quotation marks here because this parameter has, in fact, a special name in FM theory: *frequency deviation* (represented as *d*), and its value is usually expressed in Hz.

The parameters for the FM instrument portrayed in Figure 2.7 are as follows:

d = frequency deviation

f_m = modulator frequency

a_c = carrier amplitude

f_c = offset carrier frequency

The role of frequency deviation and of the modulator frequency is illustrated in Figure 2.8. Both parameters can drastically change the form of the carrier wave. If the modulator frequency is kept constant whilst increasing the frequency deviation, then the period of the carrier's output will increasingly expand and contract, proportional to the frequency deviation. If the frequency deviation remains constant and the modulator frequency is increased, then the rate of the deviation will become faster.

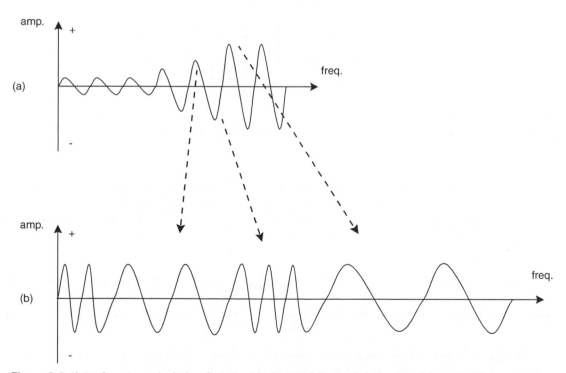

Figure 2.8 If the frequency deviation (i.e. the amplitude of the modulator signal) is gradually increased, then the period of the carrier's output increasingly expands and contracts, proportional to the value of the frequency deviation

2.2.2 The spectrum of FM sounds

The simplicity of its architecture and its capability to produce a great variety of different timbres made FM synthesis more attractive than other techniques available at the time of its invention. Moreover, it does not necessarily need waveforms other than sinusoids, for both modulator and carrier, in order to produce interesting musical sounds.

Calculating the frequencies of the partials

The spectrum of an FM sound is composed of the offset carrier frequency (f_c) and a number of partials on either side of it, spaced at a distance equal to the modulator frequency (f_m). The partials generated on each side of the carrier frequency are usually called sidebands. The sideband pairs are calculated as follows: $f_c + k \times f_m$ and $f_c - k \times f_m$ where k is an integer, greater than zero, which corresponds to the order of the partial counting from f_c (Figure 2.9).

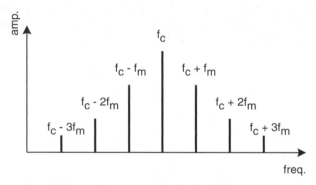

Figure 2.9 The spectrum of an FM sound is composed of the offset carrier frequency and a number of partials on either side of it, spaced at a distance equal to the modulator frequency

The amplitudes of partials are determined mostly by the frequency deviation. When there is no modulation (i.e. $d = 0$) the power of the signal resides entirely in the offset carrier frequency f_c. Increasing the value of d produces sidebands at the expense of the power in f_c. The greater the value of d, the greater the number of generated partials and, therefore, the wider the distribution of the power between the sidebands.

The FM theory provides a useful tool for the control of the number of audible sideband components and their respective amplitudes: the *modulation index*, represented as i. The modulation index is the ratio between the frequency deviation and the modulator frequency: $i = d/f_m$.

As the modulation index increases from zero, the number of audible partials also increases and the energy of the offset carrier frequency is distributed among them (Figure 2.10). The number of sideband pairs with significant amplitude can generally be predicted as $i + 1$; for example, if $i = 3$ then there will be four pairs of sidebands surrounding f_c.

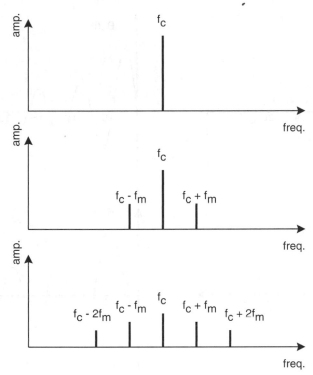

Figure 2.10 As the modulation index increases, the number of audible partials also increases and the energy of the offset carrier frequency is distributed among them

Calculating the amplitudes of the partials

The offset carrier frequency may often be the most prominent partial in an FM sound; in this case it will define the pitch of the sound. Sound engineers tend to refer to the carrier frequency as the fundamental frequency, but musicians often avoid this nomenclature because in music the term 'fundamental frequency' is normally associated with pitch and in FM the carrier frequency does not always determine the pitch of the sound.

The amplitudes of the components of the spectrum are determined by a set of functions known as *Bessel functions* and represented as $B_n(i)$. An in-depth mathematical study of Bessel functions is beyond the scope of this book; we will only introduce the basics in order to understand how they determine the amplitude of the partials of an FM-generated sound.

Figure 2.11 shows the graphical representation of four Bessel functions: $B_0(i)$, $B_1(i)$, $B_2(i)$ and $B_3(i)$, respectively. They determine amplitude scaling factors for pairs of sidebands, according to their position relative to the offset carrier's frequency. Note that these are not absolute amplitude values, but scaling factors.

The carrier amplitude (a_c) usually defines the overall loudness of the sound, but the amplitudes for individual partials are calculated by scaling the given carrier amplitude according to the factors established by the Bessel functions. For instance, $B_0(i)$ determines the

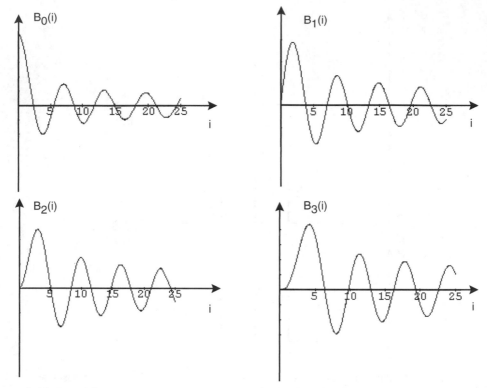

Figure 2.11 Bessel functions determine the amplitude scaling factors for pairs of sidebands, according to their position relative to the offset carrier's frequency

scaling for the offset carrier frequency (f_c), $B_1(i)$ the scaling for the first pair of sidebands ($k = 1$), $B_2(i)$ the scaling for the second pair of sidebands ($k = 2$), and so on.

The vertical axis indicates the amplitude scaling factor according to the value of the modulation index represented by the horizontal axis. For example, when $i = 0$ (i.e. no modulation) the offset carrier frequency will sound at its maximum factor (i.e. 1) and the amplitudes of all sidebands will be zero; if $i = 1$, then the scaling factor for f_c will decrease to approximately 0.76, the factor for the first pair of sidebands will have a value of 0.44, the second pair will have a value of 0.11, etc. (Figure 2.12). The value of the modulation index must be large in order to obtain significant amplitudes in high-order sidebands. See Appendix 1 for a list of scaling factors.

One important rule to bear in mind when calculating an FM spectrum is that the scaling factors for the odd partials on the left of the sideband pair are multiplied by –1.

Dealing with negative amplitudes

In addition to the above rule, notice that Bessel functions indicate that sidebands may have either positive or 'negative' amplitude, depending on the modulation index. For example, if

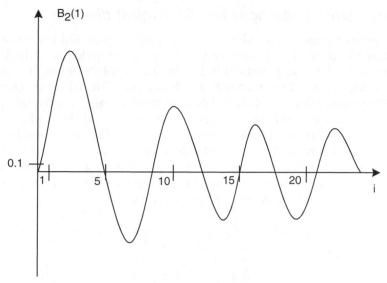

Figure 2.12 The vertical axis of a Bessel function plot indicates the amplitude scaling factor according to the value of the modulation index represented by the horizontal axis. For example, the amplitude scaling factor for the second pair of sidebands will be value 0.11 if $i = 1$; i.e. $B_2(1) = 0.11$

$i = 5$, the scaling factor for the first pair of sidebands will be approximately −0.33. In fact, 'negative' amplitude does not exist. The negative signal here indicates that the sidebands are out of phase. This may be represented graphically by plotting it downwards on the frequency axis (Figure 2.13), but the phase of a partial does not produce an audible effect unless another partial of the same frequency happens to be present. In this case, the amplitudes of the conflicting partials will either add or subtract, depending on their respective phases.

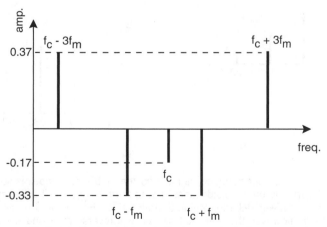

Figure 2.13 Negative amplitudes indicate sidebands that are out of phase and may be represented by plotting them downwards in the frequency axis

Dealing with negative frequencies and the Nyquist distortion

An important phenomenon to consider for the calculation of an FM spectrum occurs when the offset carrier frequency is very low and the modulation index is set high. In this case, modulation may produce sidebands that fall in the negative frequency domain of the spectrum. As a rule, all negative sidebands will fold around the 0 Hz axis and mix with the sidebands in the positive domain. Reflected partials from the negative domain will, however, reverse their phase. For instance, the settings $f_c = 440\,\text{Hz}$, $f_m = 440\,\text{Hz}$ and $i = 3$, produce the following negative sidebands: $-1320\,\text{Hz}$, $-880\,\text{Hz}$, and $-440\,\text{Hz}$. These partials fold into the positive domain, with reversed phase, and add algebraically to the partials of the same values that were there before the reflection; Figure 2.14 illustrates this phenomenon. In fact, phase information does not necessarily need to be represented in the resulting spectrum, because people do not hear the effect of phase inversion.

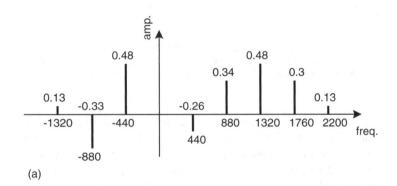

(a)

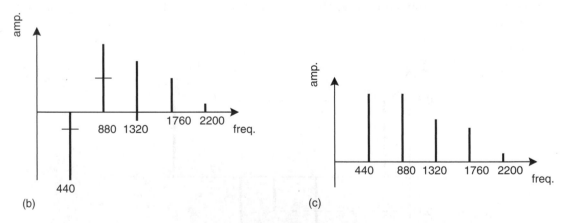

(b)   (c)

Figure 2.14 When the offset carrier frequency is low and the modulation index is set very high, the modulation will almost certainly produce sidebands that fall in the negative frequency domain of the spectrum (a). As a rule, negative sidebands will fold around the 0 Hz axis and mix with the sidebands in the positive domain. Note, however, that reflected partials will reverse their phase and add algebraically to the partials of the same values that were present before the reflection (b). Phase information does not necessarily need to be represented in the resulting plot (c)

Another phenomenon that is worth remembering here is that partials falling beyond the Nyquist limit (see Chapter 1) fold over and reflect into the lower portion of the spectrum.

2.2.3 Synthesising time-varying spectra

The ability to provide control for time-varying spectral components of a sound is of critical importance for sound synthesis. The amplitudes of the partials produced by most acoustic instruments vary through their duration. They often evolve in complicated ways, particularly during the attack of the sound. This temporal evolution of the spectrum cannot be heard explicitly at all times. Occasionally, the evolution might occur over a very short time span or the whole duration of the sound itself may be very short. Even so, it establishes an important cue for the recognition of timbre.

Frequency modulation offers an effective parameter for spectral evolution: the *modulation index* (*i*). As has been already demonstrated, the modulation index defines the number of partials in the spectrum. An envelope can thus be employed to time-vary the modulation

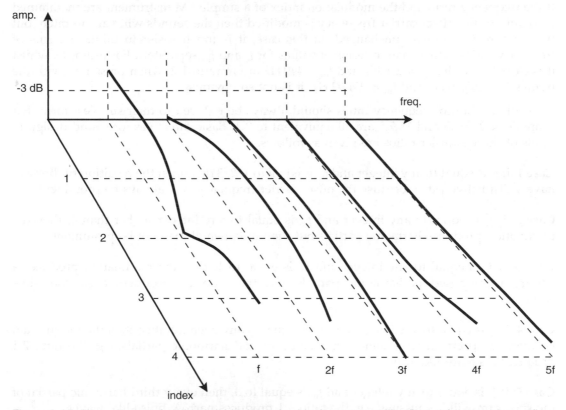

Figure 2.15 The evolution of each partial is determined by its corresponding Bessel function. A specific partial may increase or decrease its amplitude according to the slope of its Bessel function. Note, however, that linearly increasing the modulation index does not necessarily increase the amplitude of the high-order sidebands linearly

index in order to produce interesting spectral envelopes that are unique to FM. Note, however, that by linearly increasing the modulation index the instrument does not necessarily increase the power of high-order sidebands linearly.

Remember that the evolution of each partial is determined by its corresponding Bessel function. A specific partial may therefore increase or decrease its amplitude, according to the slope of its Bessel function at specific modulation values (Figure 2.15).

2.2.4 Frequency ratios and sound design

Timbre control in FM is governed by two simple ratios between FM parameters. One is the ratio between the frequency deviation and the modulator frequency and has already been introduced: it defines the modulation index (i). The other is the ratio between the offset carrier frequency and the modulator frequency, called *frequency ratio* and represented as $f_c{:}f_m$. The frequency ratio is a useful tool for the implementation of a phenomenon that is very common among conventional instruments, that is, achieving variations in pitch whilst maintaining the timbre virtually unchanged.

If the frequency ratio and the modulation index of a simple FM instrument are maintained constant but the offset carrier frequency is modified then the sounds will vary in pitch, but their timbre will remain unchanged. In this case, it is much easier to think in terms of frequency ratios rather than in terms of values for f_c and f_m separately. For example, whilst it is clear to see that $f_c = 220\,\mathrm{Hz}$ and $f_m = 440\,\mathrm{Hz}$ are in ratio 1:2, when presented with the figures $f_c = 465.96\,\mathrm{Hz}$ and $f_m = 931.92\,\mathrm{Hz}$, it is not so obvious.

As a rule of thumb, frequency ratios should always be reduced to their simplest form. For example, 4:2, 3:1.5 and 15:7.5 are all equivalent to 2:1. Basic directives for sound design in terms of these simpler ratios are given as follows:

Case 1: if f_c is equal to any integer and f_m is equal to 1, 2, 3 or 4, then the resulting timbre will have a distinctive pitch, because the offset carrier frequency will always be prominent.

Case 2: if f_c is equal to any integer and f_m is equal to any integer higher than 4, then the modulation produces harmonic partials but the fundamental may not be prominent.

Case 3: if f_c is equal to any integer and f_m is equal to 1, then the modulation produces a spectrum composed of harmonic partials; e.g. the ratio 1:1 produces a sawtooth-like wave.

Case 4: if f_c is equal to any integer and f_m is equal to any even number, then the modulation produces a spectrum with some combination of odd harmonic partials; e.g. the ratio 2:1 produces a square-like wave.

Case 5: if f_c is equal to any integer and f_m is equal to 3, then every third harmonic partial of the spectrum will be missing; e.g. the ratio 3:1 produces narrow pulse-like waves.

Case 6: if f_c is equal to any integer and f_m is not equal to an integer, then the modulation produces non-harmonic partials; e.g. 2:1.29 produces a 'metallic' bell sound.

2.2.5 Composite frequency modulation

Composite FM involves two or more carrier oscillators and/or two or more modulator oscillators. There are a number of possible combinations and each of them will create different types of spectral compositions. On the whole, complex FM produces more sidebands but the complexity of the calculations to predict the spectrum also increases. There are at least five basic combinatory schemes for building composite FM instruments:

1 Additive carriers with independent modulators
2 Additive carriers with one modulator
3 Single carrier with parallel modulators
4 Single carrier with serial modulators
5 Self-modulating carrier

Additive carriers with independent modulators

This scheme is composed of two or more simple FM instruments working in parallel (Figure 2.16). The spectrum is therefore the result of the addition of the outputs from each instrument.

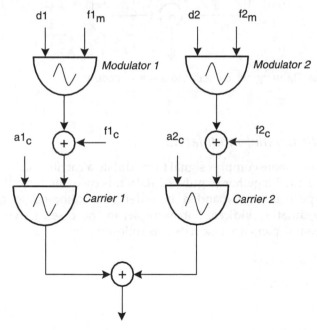

Figure 2.16 Composite FM using two carriers with independent modulators

Additive carriers with one modulator

This scheme employs one modulator oscillator to modulate two or more carrier oscillators (Figure 2.17). The resulting spectrum is the result of the addition of the outputs from each carrier oscillator.

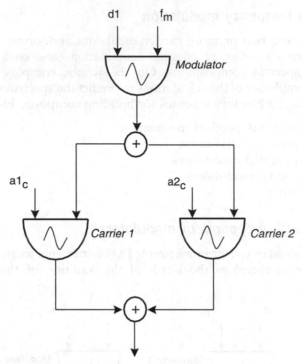

Figure 2.17 Composite FM using two carriers and a single modulator

Single carrier with parallel modulators

This scheme employs a more complex signal to modulate a carrier oscillator: the result of two or more sinewaves added together (Figure 2.18). In this case, the formula for the calculation of a simple FM spectrum is expanded in order to accommodate multiple modulator frequencies and modulation indices. For example, in the case of two parallel modulator oscillators, the sideband pairs are calculated as follows:

$$f_c - (k_1 \times f_{m1}) + (k_2 \times f_{m2})$$
$$f_c - (k_1 \times f_{m1}) - (k_2 \times f_{m2})$$
$$f_c + (k_1 \times f_{m1}) + (k_2 \times f_{m2})$$
$$f_c + (k_1 \times f_{m1}) - (k_2 \times f_{m2})$$

This formula looks complicated but, in fact, it simply states that each of the partials produced by one modulator oscillator (i.e. $k_1 \times f_{m1}$) forges a 'local carrier' for the other modulator oscillator (i.e. $k_2 \times f_{m2}$). The larger the number of parallel modulators, the greater the amount of nested 'local carriers'. The amplitude scaling factors here result from the multiplication of the respective Bessel functions: $B_n(i_1) \times B_m(i_2)$. A practical example is given in Appendix 1.

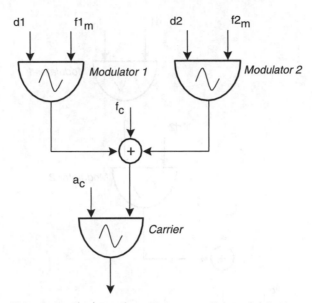

Figure 2.18 Composite FM using a single carrier with two parallel modulators

Single carrier with serial modulators

This scheme also employs a complex signal to modulate a carrier oscillator. In this case, however, the modulating signal is a frequency modulated signal (Figure 2.19). The sideband frequencies are calculated using the same method used above for parallel modulators, but the calculation of the amplitude scaling factors is different. The 'order' of the outermost modulator is used to scale the modulation index of the next modulator:

$$B_n(i_1) \times B_m(n \times i_2).$$

The main differences between the spectrum generated by serial modulators and parallel modulators, using the same frequency ratios and index, are that:

1 The former tends to have sidebands with higher amplitude values than the latter
2 No sideband components from $B_m(i)$ are generated around the carrier centre frequency; e.g. $B_0(i_1) \times B_1(0 \times i_2) = 0$

Self-modulating carrier

The self-modulating carrier scheme employs the output of a single oscillator to modulate its own frequency (Figure 2.20). The oscillator output signal is multiplied by a *feedback factor* (represented as f_b) and added to a frequency value (f_m) before it is fed back into its own frequency input; f_b may be considered here as a sort of modulation index.

This scheme will always produce a sawtooth-like waveform due to the fact that it works with a 1:1 frequency ratio by default; that is, the modulation frequency is equal to its own frequency. The amplitudes of the partials increase proportionally to f_b. Beware, however: as this parameter is very sensitive, values higher than $f_b = 2$ may lead to harsh white noise.

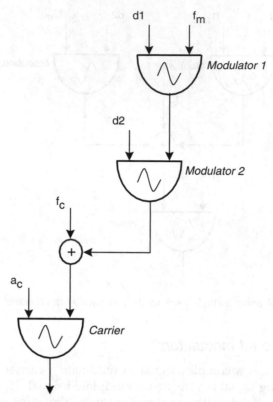

Figure 2.19 Composite FM using a single carrier with two modulators in series

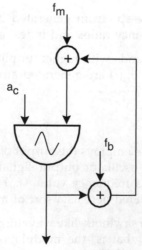

Figure 2.20 The self–modulating carrier scheme employs the output of a single oscillator to modulate its own frequency

The self-modulating carrier scheme is sometimes preferable to a simple 1:1 FM instrument. The problem with simple FM is that the amplitudes of its partials vary according to the Bessel functions, but this variation is not linear. The number of sidebands increases by augmenting the modulation index, but their amplitudes do not rise linearly. This grants an 'unnatural' coloration to the sound which may not always be desirable. The amplitudes of the partials produced by a self-modulating oscillator increase more linearly according to the feedback factor (f_b).

2.2.6 Modular implementation

There are various approaches for implementing FM synthesisers; for example, a computer programming-based approach can be substantially different from a commercial MIDI-keyboard approach. Whilst the former may have highly flexible tools for instrument design, the latter may not provide this flexibility but can be much easier to operate. In general, the industry of MIDI-keyboard type of synthesisers tends to produce better 'interfaces' than the academic programming languages, but at the expense of flexibility. Yamaha is a well-known trademark for commercially successful FM synthesisers and its success is probably due to an ingenious industrial procedure: basic FM mechanisms are encapsulated into higher-level modules called *operators*. In this case, the user only has access to a limited set of parameters to control these operators; for example, the modulation index is not explicitly available for manipulation.

An operator basically consists of an envelope generator and an oscillator, and it can be used as either a carrier operator or a modulator operator (Figure 2.21). Synthesisers are programmed by appropriately interconnecting a number of operators to form what is referred to as an *algorithm*. Various instruments are then defined by specifying appropriate parameter values for specific topologies of algorithms (Figure 2.22). An algorithm plus a set of definite parameter values constitutes a *patch*. A collection of patches is then provided in read-only memory (ROM) but the user can edit the parameter values and the algorithm in

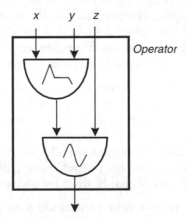

Figure 2.21 The operator of a DX7 type synthesiser basically consists of an envelope generator and an oscillator and it can be used either as a carrier or a modulator

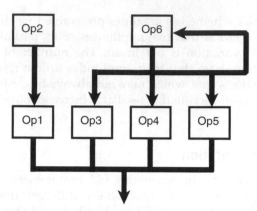

Figure 2.22 Some FM synthesisers are programmed by appropriately interconnecting a number of operators to form different algorithms

order to create his or her own patches. Depending on the model of the synthesiser, four or six operators are and can be combined into a number of different algorithms.

2.2.7 FM examples

On the accompanying CD-ROM, Kenny McAlpine's Audio Architect tutorial (in folder *audiarch*) illustrates how to implement a simple FM instrument. A number of FM examples are also available in pcmusic (in folder *Fm* in the pcmusic materials) and Nyquist (check the tutorial in the Nyquist folder). There are also a couple of FM examples in Reaktor.

2.3 Waveshaping synthesis

Waveshaping synthesis (also termed *non-linear distortion* or *non-linear processing*) creates composite spectra by applying distortion to a simple sound (e.g. a sinusoid). The technique functions by passing a sound through a unit that distorts its waveform, according to a user-specified directive. In waveshaping synthesis jargon, the distortion unit is called *waveshaper* and the user-specified directive is called *transfer function*. The first experiments with this synthesis technique were carried out by the composer Jean-Claude Risset in the late 1960s. Waveshaping has been explored substantially by Daniel Arfib and Marc Le Brun (Arfib, 1979).

As a metaphor to understand the fundamentals of waveshaping synthesis, imagine a note played on an electric stringed instrument connected to a vacuum-tube amplifier. If the volume knob of the amplifier is increased to its maximum, the vacuum-tubes will be saturated and the sound will clip and if the amplitude of the note is increased at its origin, before entering the amplifier, then the output will clip even more. If the note is a sinusoid, the louder the input, the more squared the output wave will be. If the note has a complex spectrum, then the output will be a signal blurred by distortion.

Next, assume that the instrument produces a sustained note, gradually losing amplitude, as portrayed in Figure 2.23. In this case, the output will be a sound whose amplitude remains almost constant, but the waveform will vary continuously.

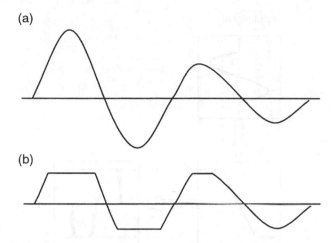

Figure 2.23 In waveshaping, the richness of the resulting spectrum (b) is proportional to the amplitude of the input signal (a)

Amplitude sensitivity is one of the key features of waveshaping synthesis. In the above example, the amount of amplification (or distortion) is proportional to the level of the input signal. The spectrum of the output therefore becomes richer as the level of the input is increased.

2.3.1 The waveshaper

The crucial element of the waveshaping technique is the *waveshaper* (Figure 2.24): a processor that can alter the shape of the waveform passing through it.

In a linear processor, such as an *ideal amplifier*, a change in the amplitude of the input signal produces a similar change in the output signal, but the shape of the waveform remains unchanged. For example, doubling the amplitude of the input signal will double the amplitude of the output. Conversely in a non-linear processor, the relationship between input and output signals depends upon the amplitude of the input signal and the nature of the non-linearity. In most cases, non-linear processors modify the waveform of the input signal. This modification generally results in an increase in the number and intensity of its partials.

The waveshaper is characterised by *a transfer function* which relates the amplitude of the signal at the output to the signal at the input. For example, Figure 2.25 shows a graphic representation of a transfer function where the amplitude of the input signal is plotted on the horizontal axis and the output on the vertical. For a given input value, the output of the waveshaper can be determined by computing the corresponding output value on the graph of the transfer function. If the transfer function is a straight diagonal line, from –1 to +1, the output is an exact replica of the input. Any deviation from a straight diagonal line introduces some sort of modification.

input signal

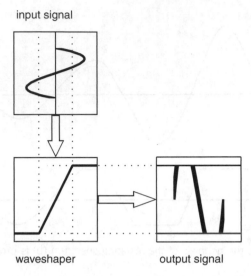

waveshaper   output signal

Figure 2.24 The waveshaper is a processor that can alter the shape (i.e. form) of the waveform passing through it

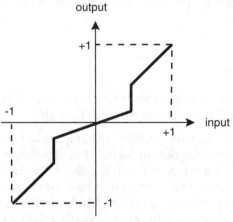

Figure 2.25 A graphic representation of a transfer function where the amplitude of the input signal is plotted on the horizontal axis and the output on the vertical

The success of a waveshaping instrument depends upon the choice of a suitable transfer function. Drawing transfer functions would intuitively be one way to experiment with waveshaping, but it might become difficult to predict the results systematically. Fortunately, there are a number of tools for constructing waveshapers out of mathematical functions that allow for some degree of prediction of the resulting spectra; for instance, *polynomials* and *trigonometric functions*. To a great extent the best transfer functions are described using polynomials. See Appendix 1 for detailed mathematical specifications.

2.3.2 Chebyshev polynomials and spectra design

A particular family of polynomials called *Chebyshev polynomials of the first kind* has been widely used for specifying transfer functions for waveshaping synthesis. Chebyshev polynomials are represented as follows: $T_k(x)$ where k represents the order of the polynomial and x represents a sinusoid. Chebyshev polynomials have the useful property that when a cosine wave with amplitude equal to one is applied to $T_k(x)$, the resulting signal is a sinewave at the kth harmonic. For example, if a sinusoid of amplitude equal to one is applied to a transfer function given by the seventh-order Chebyshev polynomial, the result will be a sinusoid at seven times the frequency of the input. Chebyshev polynomials for T_1 to T_{10} are given in Appendix 1.

Because each separate polynomial produces a particular harmonic of the input signal, a certain spectrum composed of various harmonics can be obtained by summing a weighted combination of Chebyshev polynomials, one for each desired harmonic (see Appendix 1 for an example).

Given the relative amplitudes of the required harmonics, most sound synthesis programming languages and systems provide facilities to plot Chebyshev-based transfer functions on a waveshaper. On most systems, the waveshaper is implemented as a table, or array, whose size is compatible with the maximum amplitude value of the input signal. For instance, to fully waveshape a sinusoid oscillating between −4096 and +4096 sampling units, the transfer function should span a table containing 8193 values (i.e. 4096 for the negative samples, plus 4096 for the positive, plus 1 for the zero crossing). Note that the expression 'amplitude 1' has been conventionally used on a number of occasions here to refer to the maximum input amplitude value that a waveshaper can manage.

2.3.3 Distortion index and time-varying spectra

Variations on the amplitude of the input signal can activate different 'portions' of the waveshaper. Waveshaping synthesis is therefore very convenient to synthesise sounds with a considerable amount of time-varying spectral components.

The amplitude of the input signal is often associated with a *distortion index*, controlled by an envelope. In this case, changes in volume will cause changes in the spectrum. This coincides in theory with the behaviour of a number of acoustic instruments, where louder playing produces more overtones.

2.4 Walsh synthesis

Whereas AM, FM and waveshaping are synthesis techniques inherited from the world of analog synthesis, Walsh synthesis is an inherently digital technique. Walsh synthesis works based upon square waves instead of sinewaves; it creates other waveforms by combining square waves. Inasmuch digital systems work with only two numbers at their most fundamental level (i.e. 0 and 1), square waves are very straightforward to manipulate digitally.

It is curious to note that although Walsh synthesis may resemble the additive synthesis method of sinewave summations, this technique is based upon a very different paradigm.

Walsh synthesis produces waveforms from the summation of Walsh functions rather than from the summation of partials. Walsh functions are measured in *zero crossings per second* (zps) instead of *cycles per second* (cps or Hz) and they are not explicitly related to specific 'harmonics' (Figure 2.26).

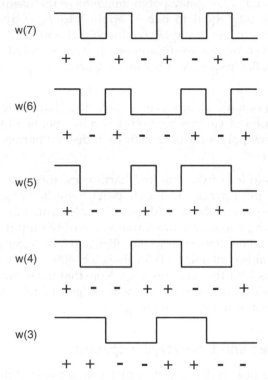

Figure 2.26 Walsh functions are the backbone of Walsh synthesis

Although is it possible to predict the spectrum of Walsh synthesis by means of complex mathematical calculations, this technique is not efficient for methods rooted in Fourier's theory. Actually, the most difficult waveform to produce with Walsh synthesis is the sinewave, which is the fundamental unit of Fourier's theory. Conversely, the most difficult waveform to produce by summing sinewaves is the square wave, which is the fundamental unit of Walsh synthesis.

Despite its affinity with the digital domain, Walsh synthesis has not been sufficiently studied and composers have rarely used it. There is great potential, however, for an entirely new theory for sound synthesis based on Walsh functions; for example, sounds could be analysed using a Walsh-based technique and resynthesised using fast Walsh transform (FWT). Also, modulation involving Walsh functions may be able to produce certain types of waveforms much more efficiently than modulation using sinusoids. A few technical articles about Walsh synthesis can be found in various issues of *Journal of the Audio Engineering Society*; e.g. Hutchins (1973, 1975).

2.5 Binary instruction

Synthesis by binary instruction is a curious technique developed in the 1970s at the Institute of Sonology in the Netherlands. This technique is commonly known as *non-standard synthesis*, but in this book we prefer to use the more specific term – binary synthesis – in order to distinguish it from other methods that could also be classified as non-standard; for example, Walsh synthesis and sequential waveform composition (discussed in Chapter 5).

Most synthesis techniques presented in this book have their roots, in one way or another, in metaphorical models seeking inspiration from the acoustic world. The great majority of systems for software synthesis actually bias the design of instruments towards this type of inspiration. The binary instruction technique diverges from this approach. It departs from the idiomatic use of low-level computer instructions, with no attempt to model anything acoustic.

Standard systems for software synthesis normally generate samples from a relatively high-level specification of acoustically related components and parameters, such as oscillators, envelope, frequency and amplitude. In binary instruction, samples are generated from the specification of low-level computer instructions with no reference to any pre-defined synthesis paradigm. The sound is described entirely in terms of digital processes. The rationale for this radical approach is that the sounds should be produced using the low-level 'idiom' of the computer. If there is such a thing as a genuine 'computer timbre', then binary instruction would be the most qualified technique to synthesise it.

Binary instruction functions by using basic computer instructions such as logical functions and binary arithmetic to process sequences or binary numbers. The result of the processing is output as samples through an appropriate digital-to-analog converter (DAC). The compelling aspect of this technique is the speed in which samples are processed. Since there is no need to compile or interpret high-level instructions, sounds can be produced in real time on very basic machines. This was a very great achievement for the 1970s.

Different sounds are associated with different programs coded in the assembler language of the computer at hand. The idea is interesting but it turned out to be a damp squib because assembler is a very low-level language and musicians would hardly have a vested interest in this level of programming. In order to alleviate the burden of writing assembler programs to produce sounds, Paul Berg and his colleagues at the Institute of Sonology developed a language called PILE (Berg, 1979) for the specification of binary instruction instruments. Similarly, at the University of Edinburgh, Stephen Holtzman developed a sophisticated system for generating binary instruction instruments inspired by research into artificial intelligence (Holtzman, 1978).

2.6 Wavetable synthesis

The term 'wavetable synthesis' alludes to a relatively broad range of synthesis methods largely used by commercial synthesisers, ranging from MIDI-keyboards to computer sound cards. Also, expressions such as *vector synthesis* and *linear arithmetic* have been brought into play by the synthesiser industry to refer to slightly different approaches to wavetable synthesis. There are at least four major distinct mechanisms that are commonly employed to implement wavetable synthesis: single wavecycle, multiple wavecycle, sampling and crossfading.

In fact, most synthesis languages and systems, especially those derived from the Music N series, function using wavetables. In order to understand this, imagine the programming of a sinewave oscillator. There are two ways to program such oscillators on a computer. One is to employ a sine function to calculate the samples one by one and output each of them immediately after their calculation. The other method is to let the machine calculate only one cycle of the wave and store the samples in its memory. In this case, the sound is produced by repeatedly scanning the samples as many times as necessary to forge the duration of the sound (Figure 2.27). The memory where the samples are stored is technically called *wavetable* or *lookup table*; hence the scope for the ambiguity we commonly encounter with this term.

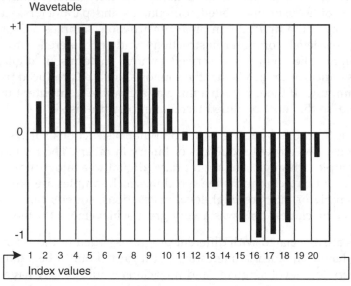

Figure 2.27 The samples for a waveform are stored on a wavetable and the sound is produced by repeatedly scanning the samples

2.6.1 Single wavecycle

The single wavecycle approach works with wavetables containing only one cycle of sounds. This is normally the case for oscillator units found in software synthesis systems; e.g. pcmusic *osc*. The result is a repetitive static sound without any variance in time (Figure 2.28). Single wavecycle-based synthesisers normally provide a bank of waveforms such as sine, triangle and sawtooth, and in some systems these waveforms can be mixed together for playback. More complex set-ups also allow for the selection of different methods for playing back the cycle; for instance, by concatenating between straight and reverse playback or even by concatenating different types of waveforms (Figure 2.29).

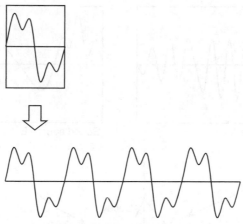

Figure 2.28 The single wavecycle approach to wavetable synthesis

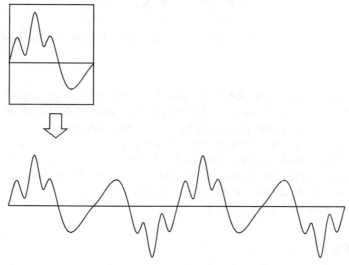

Figure 2.29 Single wavecycle concatenating between straight and reverse playback

In order to introduce some variance over time, wavecycle oscillators should be used in conjunction with signal modifiers such as envelopes and low-frequency oscillators (LFO).

2.6.2 Multiple wavecycle

The multiple wavecycle approach is similar to that of the single wavecycle. The main differences are that a wavetable may hold more than one cycle of a sound and the samples are usually from recordings of 'natural' instruments, as opposed to synthetically generated waveforms. This is the approach adopted by the manufacturers of most computer sound cards. These cards usually have a bank of recorded timbres in read-only memory (ROM) plus

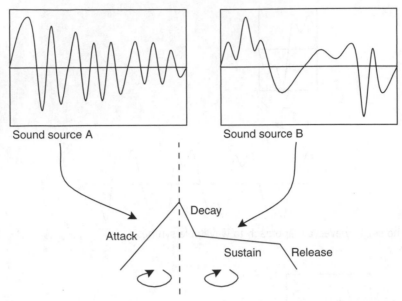

Figure 2.30 Multiple wavecycle using two different sound sources for different portions of the sound

space in random access memory (RAM) to add a few others, either sampled by the user or provided by a third party (e.g. on CD or from the Internet). Additional effects and simple sound-processing tools are often provided to allow some degree of user customisation. For example, an envelope can be redrawn or a low-pass filter can be used to dump higher partials. Other sophisticated mechanisms, such as using a sequence of partial wavecycles to form a complete sound, are commonly associated with the multiple wavecycle approach; for example, when the attack, decay, sustain and release portions of the sound are taken from different sources (Figure 2.30).

2.6.3 Sampling

In some cases it may be advantageous to use longer wavetables. The difference between multiple wavecycling and sampling is that the latter generally uses longer wavetables. Even though the multiple wavecycle approach works with more than one cycle, the size of the wavetables is relatively short and may need several loops to produce even a note of short duration. One advantage of sampling over multiple wavecycling is that longer wavetables allow for the use of pointers within a sample in order to define internal loopings (Figure 2.31). Due to this flexibility, sampling is normally used for creating sonorities and effects that would not be possible to create acoustically.

2.6.4 Crossfading

Crossfading is an approach which combines elements of multiple wavecycling and sampling. This works by gradually changing the sample outputs from one wavetable by the

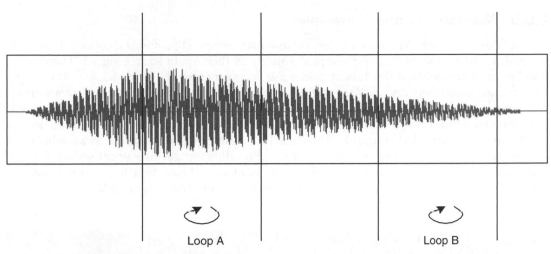

Figure 2.31 Sampling with two internal loopings

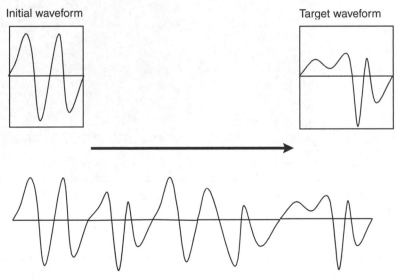

Figure 2.32 Crossfading works by gradually changing the sample outputs from one wavetable to the sample outputs of another

sample outputs of another in order to produce mutating sounds (Figure 2.32). Crossfading is a simple way of creating *sound morphing* effects, but it does not produce good results all the time. Some architectures let musicians specify more than only two sources to crossfade, and more sophisticated set-ups allow for the use of manual controllers, such as a joystick, to drive the crossfading. Variants of the crossfading approach use alternatives techniques (e.g. sample interpolation) in order to make the transition smoother.

2.6.5 Wavetable synthesis examples

Virtual Waves for MS Windows platforms (in folder *virwaves*) provides a collection of tools for making instruments that use sampled sounds as their main source signal. There are a number of instruments of this type in folder *Synthvir*; for example, *Android1.syn*, *Atomic.syn*, *Cut-up2.syn*, *Cyborg.syn* and *Zozozong.syn*. These instruments perform specific operations on the sampled sound in order to produce the desired effect; for instance, the instrument Android.syn amplitude modulates the signal in order to create extra sidebands in the spectrum of the sampled sound. In this same lot of examples, *Crossfad.syn* is an admirable case of crossfading synthesis where the sound of a cello changes to a vocal sound. But the most genuine example of wavetable synthesis is *Drawing1.syn*, which features a unique Virtual Waves synthesis module: the *Spectral Sketch Pad* module (Figure 2.33).

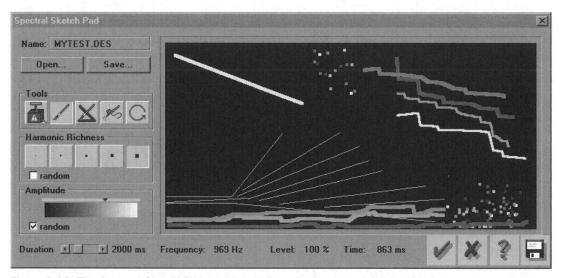

Figure 2.33 The Spectral Sketch Pad module of Virtual Waves allows for the creation of sounds graphically on a sketch pad area

This module allows for the creation of sounds graphically on a sketch pad area where the horizontal axis corresponds to time control and the vertical axis to frequency control. There are three sketching tools available:

- The paintbrush: for drawing freehand lines
- The straight line pencil: for drawing straight lines
- The airbrush: for creating patches of spots, or 'grains' of sound

3 Spectrum modelling approaches: from additive to analysis–resynthesis and formant

Until recently, the supremacy of pitch over other sound attributes prevailed in Western music theory. Time and timbre, for example, were considered to be of secondary importance. Today, however, this ideology is changing, and the tendency is now to consider that attributes such as pitch, time and timbre are all descriptors for the same phenomenon, but from different perspectives. For example, pitch may be considered as one aspect of the perception of timbre and timbre as one aspect of the perception of time. Pitch may, in turn, be considered as a quantifier to measure time. The music and theoretical work of composers such as Karlheinz Stockhausen, Pierre Schaeffer, Iannis Xenakis, Jean-Claude Risset, Horacio Vaggione, Trevor Wishart and Jorge Antunes have helped to emancipate the role of time and timbre in Western music (Stockhausen, 1959; Schaeffer, 1966; Xenakis, 1971; Risset, 1992; Vaggione, 1993; Wishart, 1985, 1994; Antunes, 1995).

Western musicians have traditionally worked with conceptual models of individual instruments, which have specific boundaries of abstraction that give very little room for the significant manipulation of timbre. In the quest for methods to define new conceptual models that consider the importance of timbre in music composition, many composers and researchers have sought inspiration from psychological theories of sound perception. This includes the outcome of two relatively new scientific fields, namely psychoacoustics and cognitive sciences (Deutch, 1982; McAdams and Deliege, 1985). In this context a new paradigm for sound synthesis has emerged naturally: *spectral modelling*. Spectral modelling techniques employ parameters that tend to describe the sound spectrum, regardless of the acoustic mechanisms that may have produced them.

Spectral modelling techniques are the legacy of the *Fourier analysis* theory. Originally developed in the nineteenth century, Fourier analysis considers that a pitched sound is made up of various sinusoidal components, where the frequencies of higher components are integral multiples of the frequency of the lowest component. The pitch of a musical note is then assumed to be determined by the lowest component, normally referred to as the *fundamental frequency*. In this case, timbre is the result of the presence of specific components and their relative amplitudes, as if it were the result of a chord over a prominently loud fundamental with notes played at different volumes. Despite the fact that not all interesting musical sounds have a clear pitch and the pitch of a sound may not necessarily correspond to the lower component of its spectrum, Fourier analysis still constitutes one of the pillars of acoustics and music.

3.1 Additive synthesis

Additive synthesis is deeply rooted in the theory of Fourier analysis. The technique assumes that any periodic waveform can be modelled as a sum of sinusoids at various amplitude envelopes and time-varying frequencies. An additive synthesiser hence functions by

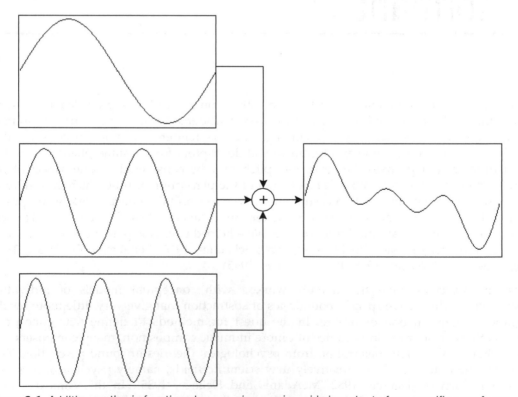

Figure 3.1 Additive synthesis functions by summing up sinusoids in order to form specific waveforms

summing up individually generated sinusoids in order to form a specific sound (Figure 3.1). Gottfried Michael Koenig was one of the first composers to create a piece entirely using additive synthesis: *'Klangfiguren I'*, composed in 1955 in Cologne, Germany.

Additive synthesis is accepted as perhaps the most powerful and flexible spectral modelling method. However, this technique is difficult to control manually and is computationally demanding. Musical timbres are composed of dozens of time-varying partials, including harmonic, non-harmonic and noise components. It would require dozens of oscillators, noise generators and envelopes to obtain convincing acoustic simulations using the additive technique. The specification and control of the parameter values for these components are also difficult and time consuming. Alternative methods have been proposed to improve this situation by providing tools to obtain the synthesis parameters automatically from the analysis of the spectrum of sampled sounds.

Although additive synthesis specifically refers to the addition of sinusoids, the idea of adding simple sounds to form complex timbres dates back to the time when people started to build pipe organs. Each pipe produced relatively simple sounds that combined to form rich spectra. In a way, it is true to say that the organ is the precursor of the synthesiser.

Robert Thompson's pcmusic tutorial on the accompanying CD-ROM (in the folder *pcmusic*) provides step-by-step instructions for building additive synthesis instruments. Also, Pedro Morales provides a number of additive synthesis examples in Nyquist (in the folder *demos*, within the Nyquist materials, in either PC- or Mac-compatible partitions of the accompanying CD-ROM): open the file *example-home.htm* using a Web browser. For an easy-to-learn programming language, specifically designed for additive synthesis, there is Som-A (in the folder *soma*). Kenny McAlpine's tutorial provides step-by-step instructions to building additive synthesis instruments in Audio Architect (in the folder *audiarch*).

3.2 An introduction to spectrum analysis

Spectrum analysis is fundamentally important for spectral modelling because samples alone do not inform the spectral constituents of a sampled sound. In order to model the spectrum of sounds, musicians need adequate means to dissect, interpret and represent them. A number of methods have been created to analyse the spectrum of sounds. There are two categories of spectrum analysis: *harmonic* and *formant*.

Whilst the former category is aimed at the identification of the frequencies and amplitudes of the spectrum components (Figure 3.2(a)), the latter uses the estimation of the overall shape of the spectrum's amplitude envelope (Figure 3.2(b)). *Short-time Fourier transform* (STFT) and *wavelet analysis* are typical examples of harmonic analysis, and *predictive analysis* is a typical example of formant analysis. Both categories have their merits and limitations; as far as sound synthesis is concerned, there is no optimum analysis technique. Some may perform better than others, according to the nature of the task at hand.

3.2.1 Short-time Fourier transform

Short-time Fourier transform (STFT) stands for an adaptation, suitable for computer programming, of the original Fourier analysis mathematics for calculating harmonic spectra.

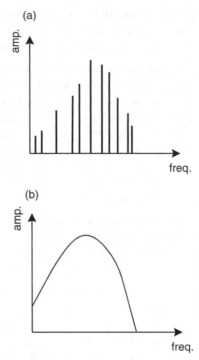

Figure 3.2 The two categories of spectrum analysis: (a) harmonic and (b) formant

It is important to bear in mind that there are at least four other terms that are closely related to STFT, which are prone to cause confusion: *Fourier transform* (FT), *discrete-time Fourier transform* (DTFT), *discrete Fourier transform* (DFT) and *fast Fourier transform* (FFT). An in-depth discussion of the differences between these terms is beyond the scope of this book. On the whole, they differ in the way that they consider time and frequency, as follows:

1 FT is the original Fourier analysis mathematics whereby time and frequency are continuous
2 DTFT is a variation of FT in which time is discrete, but frequency is continuous
3 DFT is a further version of FT in which both time and frequency are discrete
4 FFT is a faster version of DFT especially designed for computer programming

Fourier analysis detects the harmonic components of a sound using a pattern-matching method. In short, it functions by comparing a self-generated *virtual signal* with an input signal in order to determine which of the components of the former are also present in the latter. Imagine a mechanism whereby the components of the input signal are scanned by multiplying it by a reference signal. For instance, if both signals are two identical sinewaves of 110 Hz each, then the result of the multiplication will be a sinusoid of 220 Hz, but entirely offset to the positive domain (Figure 3.3). The offset value depends upon the amplitudes of both signals. Thus, it is also possible to estimate the amplitude of the input signal by taking the amplitude of the virtual signal as a reference.

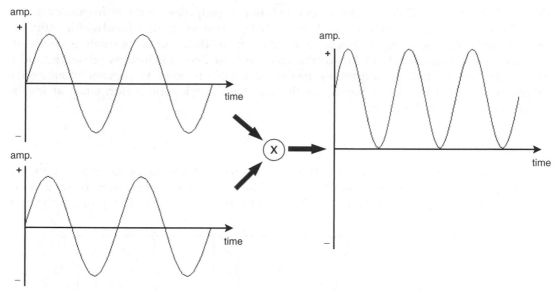

Figure 3.3 The result of the multiplication of two identical sinusoid signals will be a sinusoid entirely offset to the positive domain

The mathematics of the Fourier transform suggest that the harmonics of a composite signal can be identified by the occurrences of matching frequencies, whilst varying the frequency of the reference sinewave continuously. The digital version of this method may be simulated on a computer by scanning the input signal at rates that are multiples of its own fundamental frequency. This is basically how FFT works.

One of the main problems with the original Fourier transform theory is that it does not take into account that the components of a sound spectrum vary substantially during its course. In this case, the result of the analysis of a sound of, for example, 5 minutes' duration would inform the various components of the spectrum but would not inform when and how they developed in time. For instance, the analysis of a sound that starts as a flute and changes its timbre to a violin during its course would only display the existence of the components that form both timbres, as if they were heard simultaneously. Short-time Fourier transform implements a solution for this problem. It chops the sound into short segments called windows and analyses each segment sequentially. It normally uses FFT to analyse these windows, and plots the analysis of the individual windows in sequence in order to trace the time evolution of the sound. The result of each window analysis is called an *FFT frame*. Each FFT frame contains two types of information: a *magnitude spectrum* depicting the amplitudes of every analysed component and a *phase spectrum* showing the initial phase for every frequency component.

In order to understand the functioning of the FFT mechanism, imagine a filtering process in which a bank of *harmonic detectors* (e.g. band-pass filters) are tuned to a number of frequencies (refer to Chapter 4 for a brief introduction to filters). The greater the number of harmonic detectors, the more precise the analysis. For instance, if each detector has a bandwidth of 30 Hz, it would be necessary to employ no less than 500 units to cover a band ranging from 30 Hz to 15 000 Hz. Such a bandwidth is appropriate for lower-frequency bands where the value of a semitone is close to 30 Hz. However, this bandwidth value is unnecessarily narrow at higher-frequency bands, where 30 Hz can be as small as one tenth of a semitone. Unfortunately, FFT distributes its harmonic detectors linearly across the audio range. This means that a compromise has to be found in order to prevent unnecessary precision at higher-frequency bands, on the one hand, and poor performance at lower-frequency bands, on the other.

Windowing

In the context of STFT, windowing is the process whereby shorter portions of a sampled sound are detached for FFT analysis (Figure 3.4). The windowing process must be sequential, but the windows may overlap (Figure 3.5). The effectiveness of the STFT process depends

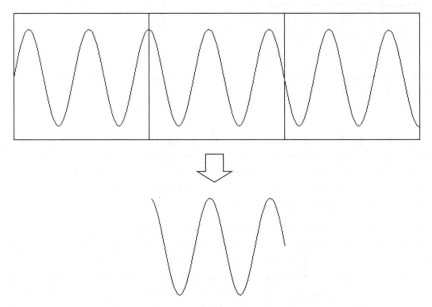

Figure 3.4 Windowing is a process in STFT whereby shorter portions of the sound are detached for FFT analysis

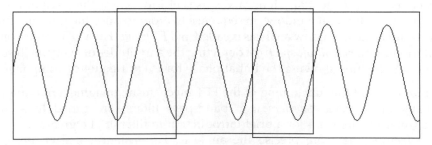

Figure 3.5 Overlapping windows

upon the specification of three windowing factors: the *envelope for the window,* the *size of the window* and the *overlapping factor.*

Note in Figure 3.4 that the windowing process may cut the sound at non-zero parts of the waveform. The FFT algorithm considers each window as a unit similar to a wavecycle. The problem with this is that interruptions between the ends of the windowed portion lead to irregularities in the analysis (Figure 3.6). This problem can be remedied by using a lobe envelope to smooth both sides of the window (Figure 3.7). From the various functions that generate lobe envelopes, the *Gaussian,* the *Hamming* and the *Kaiser* functions are more often used because they tend to produce good results.

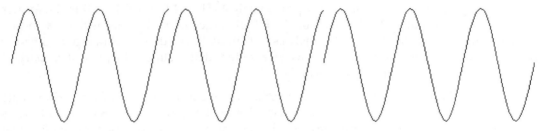

Figure 3.6 Cutting the sound at non-zero parts causes irregularities in the analysis

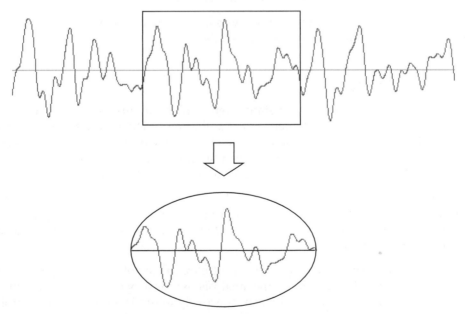

Figure 3.7 Interruptions between the end parts of the windowed portion can be remedied by using a lobe envelope to smooth both sides of the window

Lobe envelopes, however, have their own unwanted side-effects. Note that the amplitude of the whole signal now rises and falls periodically. This is a clear case of *ring modulation* (see Chapter 2) in which the 'windowing frequency' acts as the modulating frequency. Enveloped windowing also may distort the analysis process, but the result is still more accurate than with windowing without envelopes.

The size of the window defines the frequency resolution and the time resolution of the analysis. This value is normally specified as a power of two; e.g. 256, 512, 1024, etc. Longer windows have better frequency resolution than smaller ones, but the latter have better time resolution than the former. For example, whilst a window of 1024 samples at a rate of 44 100 Hz, allowing for a time resolution of approximately 23 milliseconds (1024/44 100 = 0.023), a window of 256 samples gives a much better resolution of approximately 6 milliseconds (256/44 100 = 0.0058). Conversely, the harmonic detectors will be tuned to scan frequencies spaced by a bandwidth of approximately 43 Hz (44 100/1024 = 43) in the former case and to approximately 172 Hz (44 100/256 = 172) in the latter. This means that a window of 256 samples is not suitable for the analysis of sounds lower than 172 Hz (approximately an F3 note), but it may suit the analysis of a sound that is likely to present important fluctuation within less than 23 milliseconds.

To summarise, in order to find out precisely when an event occurs, the FFT algorithm cuts down the frequency resolution and vice versa. The overlapping of successive windows is normally employed to alleviate this problem. For example, if an overlap factor is set to equal 16 (i.e. 1/16th of the size of the window) and the window size is set to equal 1024 samples, then the windowing process will slice the sound in steps of 64 samples (i.e. 1024/16 = 64). In this case the time resolution of a window of 1024 samples will improve from 23 milliseconds to approximately 1.4 milliseconds (i.e. 0.023/16 = 0.0014).

There is a tutorial on implementing FFT analysis in Nyquist on the accompanying CD-ROM (in the folder *demos*) within the Nyquist materials.

3.2.2 Wavelets analysis

The size of the STFT window is constant. Hence all harmonic detectors have the same bandwidth and are placed linearly across the audio range. The wavelets method improves this situation by introducing a mechanism whereby the size of the window varies according to the frequency being analysed. That is, the bandwidths of the harmonic detectors vary with frequency and they are placed logarithmically across the audio range.

The wavelets method is inspired by a concept known as *constant Q filtering*. In filters jargon, Q denotes the ratio between the band-pass filter's resonance centre frequency and its bandwidth. If Q is kept constant, then the bandwidths of the BPF bank vary according to their resonance centre frequency.

As far as wavelets analysis is concerned, the centre frequency of a BPF compares to the size of the window. Thus the length of the analysis window varies proportionally to the frequencies being analysed. This mechanism effectively minimises the resolution problem of the STFT analysis because it intensifies both time resolution at high-frequency bands and frequency resolution at low-frequency bands.

3.2.3 Predictive analysis

Harmonic analysis methods, such as STFT and wavelets, are not entirely adequate for the analysis of sounds with a high proportion of non-sinusoidal components and, to a certain extent, to non-harmonic combinations of partials. The nature of these signals is not compatible with the notion that sounds are composed of harmonically related and stable sinusoids. Formant analysis proposes an alternative method of representation. The sound is represented here in terms of an overall predictive mould that shapes a signal rich in partials, such as a pulse wave or white noise. The advantage of this method is that the predictive mould does not need to specify the frequencies of the spectrum precisely; any value within a certain range may qualify.

Predictive analysis is a typical example of formant analysis. The core of this method is the interplay between two kinds of filters: the *all-pole filter* and the *all-zero filter*. In electronic engineering jargon, the pole of a filter refers to a point of resonance in the spectrum and the zero of a filter to a point of attenuation. An all-pole filter is a filter that allows for several resonant peaks in the spectrum. Conversely, an all-zero filter creates various notches in the spectrum.

The main target of predictive analysis is to create an all-zero filter that corresponds to the inverse of the acoustic mould that originally shaped the spectrum of the sound in question. The analysis algorithm first reads snapshots of samples and estimates time-varying coefficients for an all-pole filter that could recreate them. These estimated coefficients are then inverted to fit an all-zero filter. Next, the all-zero filter is applied to the signal in order to test the accuracy of the estimated parameters. In theory, the all-zero filter should cancel the effect of the acoustic mould; i.e. the result of the all-zero filtering should be equivalent to the stimulation signal. The accuracy of the all-zero filter is measured by comparing its output with the original stimulation. This process is repeated until all samples of the sound have been analysed.

The main criticism of predictive analysis is that the original stimulation signal is rarely known. In practice, the algorithm predicts that the stimulation is either a pitched pulse train or white noise. In other words, the outcome of the all-zero filter is compared to an artificial stimulation that is rather rudimentary, thus limiting the efficiency of technique to specific types of sounds; e.g. prediction analysis works satisfactorily for speech and some wind sounds.

3.3 Analysis and resynthesis

Most spectral modelling techniques work in two stages – analysis and resynthesis – which can be performed in a number of ways. As has already been demonstrated, the analysis stage extracts a number of parameters from a sampled sound. The resynthesis stage then uses these parameters to recreate the sound using a suitable synthesis technique. The ability to create a precise imitation of the original sound is not, however, very interesting for composers who would obviously prefer to modify the analysis data in order to synthesise variants of the original sound. The great advantage of the analysis and resynthesis technique over plain sampling is that analysis data can be modified in a variety of ways in order to create new sounds. Typical modifications include spectral time stretching (Figure 3.8), spectral shift (Figure 3.9) and spectral stretching (Figure 3.10).

3.3.1 Resynthesis by Fourier analysis reversion

The resynthesis by Fourier analysis reversion method (sometimes referred to as *overlap-add* resynthesis) takes the information from each frame of the STFT analysis and recreates an approximation of the original sound window by window. This process works well for those

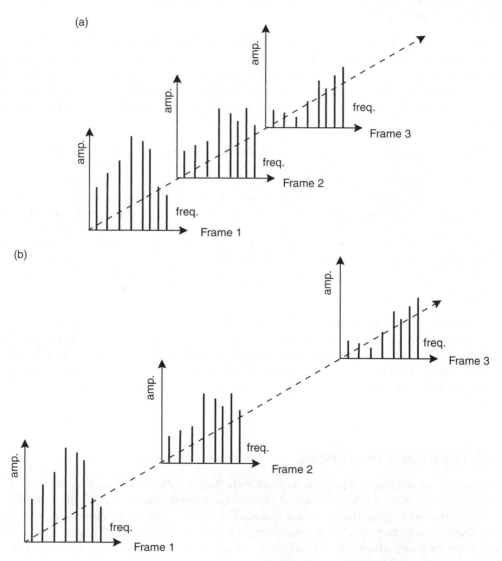

Figure 3.8 Spectral time stretching works by enlarging the distances between the frames of the original sound (a) in order to stretch the spectrum of the sound (b)

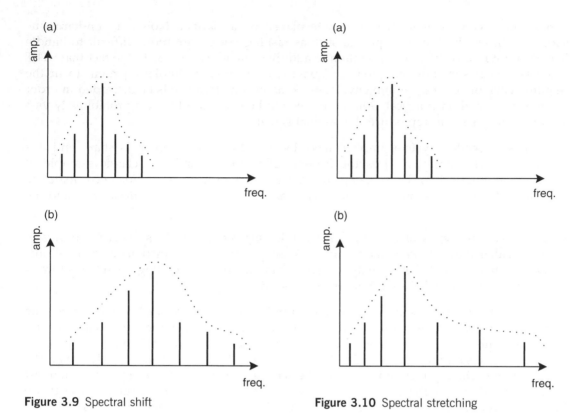

Figure 3.9 Spectral shift   **Figure 3.10** Spectral stretching

cases where the windows are reconstructed in conformity with the original analysis specification (e.g. use of similar windowing and overlapping). The analysis data must be cautiously manipulated in order to produce good results, otherwise the integrity of the signal may be completely destroyed. This may, of course, be desirable for some specific purposes, but in most circumstances such spurious effects are difficult to control. This resynthesis method is of limited interest from the standpoint of a composer who may require greater flexibility to transform the sound.

3.3.2 Additive resynthesis

This method employs STFT analysis data to control an additive synthesiser. In this case, the analysis stage includes an algorithm that converts the STFT data into amplitude and frequency trajectories – or envelopes – that span across the STFT frames. This is generally referred to as the *peak-tracking algorithm* and can be implemented in a number of ways.

The additive resynthesis method is more flexible for parametric transformation than the analysis reversion technique discussed above, because envelope data are generally straightforward to manipulate. Transformations such as stretching, shrinking, rescaling and shifting can be successfully applied to either or both frequency and time envelopes. This resynthesis method is suitable for dealing with harmonic sounds with smoothly changing, or almost static, components, such as a note played on a clarinet. Noisy or non-harmonic sounds with highly dynamic spectra, such as rasping voices, are more difficult to handle here, but not impossible. This limitation of additive resynthesis is due to the fact that STFT analysis data is somehow adapted (or 'quantised' in music technology jargon) to fit the requirements for conversion into envelopes. Some information needs be discarded in order to render the method practical, otherwise it would be necessary to use a prohibitively vast additive synthesiser to reproduce the original signal.

The Phase Vocoder system implemented by the composer Trevor Wishart and his collaborators at Composer's Desktop Project (CDP) in England is a classic example of additive resynthesis. In order to operate the Phase Vocoder the musician first needs to define a number of parameters, including the *envelope for the window*, the *size of the window* and the *overlap factor*.

From the various types of envelopes for windowing available, Phase Vocoder normally employs either the *Hamming function* or the *Kaiser function*. Both functions have a bell-like shape. One may provide better analysis results than the other, depending upon the nature of the sound analysed and the kind of result expected.

The size of the window defines the number of input samples to be analysed at a time. The larger the window, the greater the number of channels, but the lower the time resolution, and vice versa. This should be set large enough to capture four periods of the lowest frequency of interest. The sampling rate divided by the number of channels should be less than the lowest pitch in the input sound. This may be set in terms of the *lowest frequency* of interest or in terms of the *number of channels required*. Note that the term 'channel' is the Phase Vocoder jargon for what we referred earlier to as harmonic detectors (Figure 3.11).

The overlap factor is the number of samples that the algorithm skips along the input samples each time it takes a new window (Figure 3.12). This factor is specified as a fraction of the window size; e.g. an eighth of the window size.

The real power of the CDP's Phase Vocoder package is the great number of tools available for spectral manipulation techniques (Wishart, 1994). There are currently over 60 tools in the package for tasks such as:

- Blurring the spectrum
- Amplifying or attenuating the whole spectrum
- Interpolating (morphing) different sounds
- Shuffling analysis windows
- Stretching the spectrum
- Imposing the spectral envelope from one sound onto another

On the accompanying CD-ROM, in folder *cdpdemo* (PC-compatible partition) there are a few programs from the CDP Phase Vocoder suite to perform the tasks listed above: CDPVOC, BLUR, GAIN, MORPH, SHUFFLE, STRETCH and VOCODE.

CDPVOC is the actual program that performs the task of analysis and resynthesis; the remaining are programs for spectral manipulations. Note that the spectral manipulation programs do not process a sound directly but the information generated by the analysis

(a)

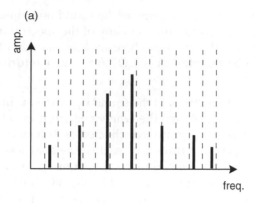

(b)

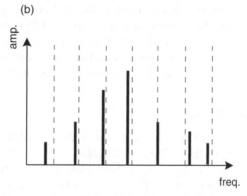

Figure 3.11 The Phase Vocoder uses channels to act as 'harmonic detectors'. The greater the number of channels, the better the frequency resolution. Note that (b) does not distinguish between the two uppermost partials that are perfectly distinct in (a)

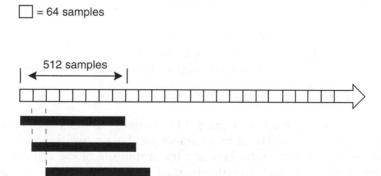

Figure 3.12 In this case a window has 512 samples and the overlap factor is set to 64 samples (512/8 = 64)

process. In order to manipulate the spectrum of the sound one must generate its analysis file first. The analysis file can then be submitted to any of the spectral transformation programs. The transformation process will always produce another analysis file, which in turn must be resynthesised in order to be heard; that is CDPVOC must return to the scene in order to resynthesise the sound.

The BLUR program blurs the spectrum of the sound by time-varying its components and the GAIN program amplifies or attenuates the whole spectrum envelope according to a user-specified factor. MORPH is a program which creates transitions between two sounds. SHUFFLE, as its name suggests, shuffles the order of the analysis windows. The STRETCH program is used to stretch the frequencies in the spectrum; that is, it expands or contracts the distances between the partials (Figure 3.10). Finally, VOCODE reshapes a sound with the characteristics of another sound, by imposing the spectral contour of the latter onto the former. Refer to the documentation on the CD-ROM for more details of the operation of these programs and for some examples. Also on the CD-ROM there are demonstration versions of Diphone (for Macintosh, in the folder *diphone*) and NI-Spektral Delay (for both PC-compatible under Windows and Macintosh platforms, in the folder *native*). Whilst Diphone is a rather sophisticated program for making transitions between sounds, NI-Spektral Delay is a fantastic spectral processor that can be controlled in real-time via MIDI.

3.3.3 Subtractive resynthesis

Subtractive resynthesis takes a rather different route from additive resynthesis. It employs formant analysis to produce the filter coefficients for a subtractive synthesiser. The advantage of the subtractive paradigm over the additive one is that it considers pitch and spectrum independently of each other. That is, in subtractive synthesis the shape of the spectrum is not subjected to a fundamental frequency, in the sense that the pitch of the stimulation signal does not affect the resonator. Note, however, that there is no guarantee that the stimulation signal will contain the components required to resonate.

Linear predictive coding (LPC) is a typical example of subtractive resynthesis employing predictive analysis. The result of the predictive analysis stage is a series of snapshots, called *LPC frames*, which contain the information required to resynthesise the sound. This information may vary in different implementations of the LPC algorithm, but it generally includes the following:

- The length of the frame (in samples or seconds)
- The average amplitude of the frame's original signal
- The average amplitude of the estimated output from the inverted filter
- The pitch for the frame
- The coefficients for the all-pole filter

The synthesis stage is depicted in Figure 3.13. Frames are read sequentially and the algorithm first decides whether the stimulation should be a noise signal or a pulse stream. This decision depends upon the ratio between the amplitude of the original signal and the amplitude of the estimated output from the inverted filter. A large ratio (e.g. higher that 0.25) leads to a choice for noise stimulation. The stimulation is then scaled to the amplitude required and fed into the all-pole filter, which may have as many as sixty poles.

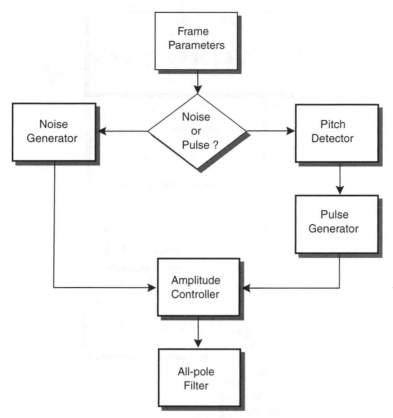

Figure 3.13 In LPC synthesis, analysis frames are read sequentially and the algorithm has to decide whether the stimulation should be a noise signal or a pulse stream

The LPC analysis frames may, of course, be edited in order to transform the sound. For instance, the duration of the frames may be stretched or shrunk in order to alter the duration of the sound without changing its pitch.

3.3.4 Resynthesis by reintegration of discarded components

This method attempts to improve additive resynthesis by adding a mechanism that reintegrates the information discarded during the STFT analysis. Each frame of the STFT is resynthesised immediately after the analysis, using the additive resynthesis method described above (Figure 3.14). This signal is then subtracted from the original in order to obtain the *residual* part that has been discarded or converted into stationary vibrations during the analysis process. This residual part is then represented as an envelope for a time-varying filter.

The outcome of the analysis process provides two branches of information: about the *sinusoidal components* of the sound and about *noisy and other non-sinusoidal components*. Note that this resynthesis method takes advantage of both categories of analysis discussed earlier:

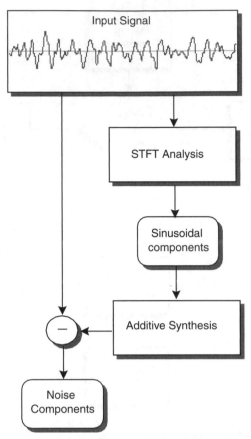

Figure 3.14 The algorithm deduces the residual part of the sound by subtracting its sinusoidal components

harmonic analysis and formant analysis. The type of representation generated by a formant analysis technique is highly suitable for representing the stochastic nature of the residual part.

The resynthesis process results from two simultaneous synthesis processes: one for sinusoidal components and the other for the noisy components of the sound (Figure 3.15). The sinusoidal components are produced by generating sinewaves dictated by the amplitude and frequency trajectories of the harmonic analysis, as with *additive resynthesis*. Similarly, the 'stochastic' components are produced by filtering a white noise signal, according to the envelope produced by the formant analysis, as with *subtractive synthesis*. Some implementations, such as the SMS system discussed below, generate 'artificial' magnitude and phase information in order to use the Fourier analysis reversion technique to resynthesise the stochastic part.

The Spectrum Modelling Synthesis system (SMS) developed by Xavier Serra, of Pompeu Fabra University, in Barcelona, Spain, is an edifying example of resynthesis by reintegration

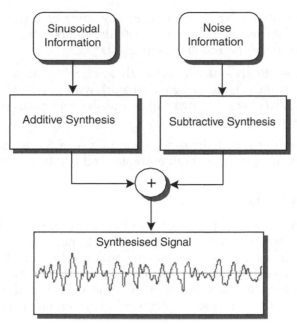

Figure 3.15 The resynthesis by reintegration of discarded components works by employing two simultaneous synthesis processes: one for the sinusoidal components and the other for the noisy part

of discarded components (Serra and Smith, 1990). The system package provides a number of modules and tools, including an analysis module and a synthesis module. The latter embeds facilities to modify the analysis data for resynthesis.

The analysis module has been designed to cope with a variety of different sounds, but it needs to be customised to the specific sound in question. In order to obtain a good analysis the musician must specify a number of parameters very carefully, such as:

- The format of the analysis; i.e. the algorithm is 'tuned' to target either a harmonic or a non-harmonic sound
- The STFT parameters such as window size and overlap factor
- The peak detection and peak continuation parameters such as the lowest and highest frequencies to be detected, and the spectral magnitude threshold
- Stochastic analysis parameters such as the maximum number of breakpoints for the formant envelope

The synthesis module provides a number of controllers to transform the analysis data for resynthesis. Here the musician can specify a large range of spectral transformations, such as:

- To scale the amplitude of the overall sound, specific areas of the spectrum or individual components
- To scale the frequencies of the overall spectrum, specific areas of the spectrum or individual components

- To modify the distribution of the harmonics by stretching or compressing them
- To amplitude modulate and frequency modulate the components of the spectrum
- To change the envelope of the residual components

Powerful and flexible synthesis systems are not always easy to master and the SMS system is no exception to this rule. It is fundamental to work with good analysis data in order to produce interesting sounds here. The two main requisites for effectively operating the SMS program are patience and experience.

On the CD-ROM (in folder *sms*) there is a version of Xavier Serra's SMS package for PC-compatible under Windows enriched with examples and comprehensive documentation.

3.4 Formant synthesis

Formant synthesis originated from research into human voice simulation, but its methods have been adapted to model other types of sounds. Human voice simulation is of great interest to the industrial, scientific and artistic communities alike. In telecommunications, for example, speech synthesis is important because it is comparatively inexpensive and also safer to transmit synthesis parameters rather than the voice signal. In fact, music sound synthesis owes much of its techniques to research in telecommunications systems.

In music research, voice simulation is important because the human ability to distinguish between different timbres is closely related to our capacity to recognise vocal sounds, especially vowels. Since the nineteenth century, a bulk of scientific studies has linked speech and timbre perception, from Helmholtz (1885) to current research in cognitive science (McAdams and Deliege, 1985).

The desired spectrum envelope of human singing has the appearance of a pattern of 'hills and valleys', technically called formants (Figure 3.16). The mouth, nose and throat function as a resonating tube whereby particular frequency bands are emphasised and others are attenuated. This resonating system can be simulated by a composition of parallel band-pass filters (BPF), in which each band-pass centre frequency is tuned to produce a formant peak. In the context of spectral modelling, however, formant synthesis does not attempt to build such a model. The core purpose of formant synthesis is to produce the formants using dedicated generators in which musicians specify the spectrum of the required sound, rather than the mechanism that produces it. One of the main advantages of this approach is that generators specifically designed to produce formants require less computation than their filter counterparts.

Figure 3.16 The desired spectrum envelope of human singing has the appearance of a pattern of 'hills and valleys', technically called formants

3.4.1 FOF generators

One of the most successful formant generators currently available is the FOF generator devised in the early 1980s by Xavier Rodet at Ircam, in Paris. In fact, 'FOF' is an acronym for 'Fonctions d'Onde Formantique' (Rodet, 1984). Briefly, a single FOF unit replaces the subtractive synthesis pair *stimulation function* followed by a *band-pass filter*. The signal produced by the FOF generator is equivalent to the signal generated by a pulse-driven filter.

The generator works by producing a sequence of dampened sinewave bursts. A single note contains a number of these bursts. Each burst has its own local envelope with either a steep or a smooth attack, and an exponential decay curve. The formant is the result of this local envelope. As the duration of each FOF burst lasts for only a few milliseconds, the envelope produces sidebands around the sinewave, as in *amplitude modulation*. Note that this mechanism resembles the granular synthesis technique, with the difference that the envelope of the 'FOF grain' was especially designed to facilitate the production of formants.

The FOF generator is controlled by a number of parameters, including amplitude, frequency and local envelope. The only difficulty with the FOF generator is to relate the shape of the envelope to the required formant (Figure 3.17). There are four formant parameters and two of them refer to the attack and decay of the local envelope:

- Formant centre frequency
- Formant amplitude
- Rise time of the local envelope
- Decay time of the local envelope

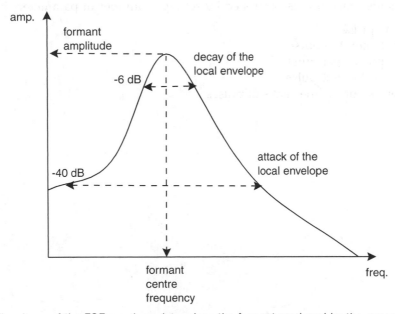

Figure 3.17 The shape of the FOF envelope determines the formant produced by the generator

The decay time local envelope defines the bandwidth of the formant at −6 dB and the rise time the skirtwidth of the formant at −40 dB. The relationships are as follows: the longer the decay, the sharper the resonance peak, and the longer the rise time, the narrower the skirtwidth. Reference values for singing sounds are approximately 7 milliseconds for the rise time and 3 milliseconds for the decay time.

This technique was originally designed for a synthesis system developed by Ircam in Paris, called *Chant* (in French 'chant' means sing). Ircam's Diphone system (in the folder *diphone* on the accompanying CD-ROM) provides a 'Chant' module whereby the composer can either specify the FOF parameters manually or infer them automatically from the analysis of given samples. Also, Virtual Waves for PC-compatible under Windows (in the folder *virwaves*) provides an FOF unit generator as part of its repertoire of synthesis modules.

3.4.2 Vosim synthesis

Vosim is an abbreviation of voice simulator. This synthesis technique was developed in the 1970s by Werner Kaegi and Stan Tempellars (1978) at the Institute of Sonology, in Utrecht, Holland. In plain terms, this technique is akin to FOF, in the sense that it produces a formant by generating sequences of tone bursts. The main difference lies in the waveform of the bursts.

A Vosim unit generates a series of bursts containing a number of pulses decreasing in amplitude. Each pulse is the result of the square of a sinewave (i.e. $\sin^2$). The *width* of each pulse determines the *centre frequency of the formant* and the end of each burst is marked by a gap which determines the *frequency of the burst*.

As with FOF, it is necessary to use several Vosim units in order to create a spectrum with various formants. Each unit is controlled by setting a number of parameters including:

- Width of the pulse
- Gap size between the bursts
- Number of pulses per burst
- Amplitude of the first pulse
- Attenuation factor for the series of pulses

4 Source modelling approach: from subtractive and waveguides to physical and modal

In contrast to the spectrum modelling approach discussed in the previous chapter, the source modelling approach focuses on the functioning of sound-producing devices, rather than on the sounds themselves. The functioning of musical instruments can be emulated in a variety of ways. The classic method starts from a system of equations describing the acoustic behaviour of an instrument when it is played. This method is appealing in principle, but it is not yet very practical for musicians: the equations are often too complex to design and they are very demanding in terms of computational resources. Moreover, these emulations often require the specification of a large amount of control information. A variety of alternatives has been proposed to alleviate this difficulty; for example, the encapsulation of these equations into functional, modular blocks. In this case, one builds an instrument by assembling these blocks. Other alternative methods employ filtering techniques and recirculating wavetables to bypass or simplify these equations.

The rationale for the desirability of source modelling is that it offers the potential to produce convincing emulations of most acoustic instruments, including the production of expressive sound attributes such as breathing and fingers sliding on a string. Moreover, this approach paves the way for more adventurous designs, such as virtual instruments that would otherwise be impossible to build; for example, a morphing didgeridoo-like instrument that shrinks to the size of a flute, changing its material from wood to metal, during the course of a melody. One should bear in mind, however, that not all the techniques introduced in this chapter are capable of fulfilling every aspect of our imagination. Some techniques may excel for certain modelling aspects but may completely fail for others.

The source modelling ideal is not new. It was used to study the behaviour of musical instruments long before electronic sound synthesis could be envisaged. Some would even argue that source modelling dates back to the early days of Western science when Pythagoras designed the monochord: a model to study music theory.

The concepts behind most of the techniques introduced in this chapter can be traced to nineteenth-century scientific treatises on the nature of vibrating objects; for example, *The Theory of Sound*, first published in 1877, by J. W. S. Rayleigh, a Fellow of Trinity College in Cambridge, England. Treatises on vibrating systems often describe the behaviour of musical instruments with the aid of mechanico-acoustic models. It was, however, the emergence of electrical models that fostered the development of electronic synthesisers and keyboards. Electrical circuits can be described by equations that are analogous to the equations used to describe mechanical and acoustic vibrating systems. Electrical models were often used to study the emulation of the acoustic behaviour of musical instruments because electrical engineering had developed powerful tools for circuitry problem solving that could be applied to the solution of vibration problems in acoustics.

It is known that various electrical models of musical instruments were created as early as the 1930s. At the beginning of the 1950s, for example, H. F. Olson (1952), of RCA Laboratories in Princeton, USA, published a book entitled *Musical Engineering*, containing various examples of the electrical modelling of musical instruments. In this book, Olson illustrates how the study of mechanico-acoustic systems is facilitated by the introduction of elements analogous to the elements in an electrical system (Figure 4.1).

Progress in turning these models into usable prototypes for music making was slow until computer technology became available. Computer modelling of vibrating systems is normally based upon the mechanical and not upon the electrical paradigm.

However, an accurate model of a musical instrument requires a significant amount of computation. Various alternatives have been designed to capture the complex behaviour of acoustic vibrating systems without requiring heavy computation; an excellent example of such alternatives is the waveguide filtering technique (Smith, 1992). The emergence of such alternatives, combined with state-of-the-art micro-chip technology enabled the industry to

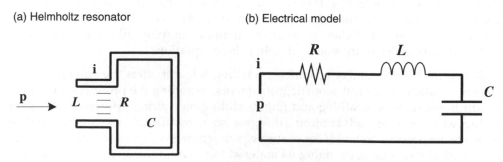

Figure 4.1 Electrical circuitry (b) may be used to model mechanico–acoustic systems (a)

release the first samples of commercial synthesisers entirely based upon the source modelling paradigm at the beginning of the 1990s: Bontempi-Farfisa's MARS in 1992 followed by Yamaha's VL1 in 1993.

4.1 Subtractive synthesis

Most musical instruments can be modelled as a resonating chamber stimulated by acoustic waveforms with certain spectral and temporal properties. Subtractive synthesis is based upon the principle that the behaviour of an instrument is determined by two main components: *excitation source* (or simply *excitation*) and *resonator*. The role of the excitation component is to produce a raw signal that will be shaped by the resonator (Figure 4.2). A variety of spectra can be obtained by varying the acoustic properties of the resonator. This technique is termed subtractive, because the filters alter the spectrum of the excitation signal by subtracing unwanted partials of its spectrum while favouring the resonation of others. For example, air blown past a cylindrical tube will produce a different sound quality, according to the dimensions of the tube.

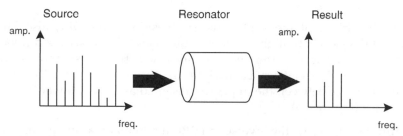

Figure 4.2 A variety of spectra can be obtained by varying the acoustic properties of the resonator

From a signal processing point of view, the resonator acts as a filter (or bank of filters) applied to a non-linear (i.e. noisy) excitation signal. Emulation of various sound spectra is achieved by combining the production of suitable source signals with the specification of appropriate coefficients for the filters.

The implementation of a subtractive synthesiser generally uses a signal generator with a rich spectrum, to act as an excitation source, and a bank of filters to simulate a resonating chamber. Two types of signal generators are commonly used as excitation sources for subtractive synthesis, because of the richness (i.e. in terms of variety of partials) of their output: *noise* and *pulse* generators. Whilst the noise generator produces a large number of random partials within a broad bandwidth, the pulse generator produces a periodic waveform, at a specific frequency, with high energy partials. The spectrum of the waveform of the pulse is determined by the ratio of the pulse width to the period of the signal; the smaller the ratio, the narrower the pulse and therefore the higher the energy of the high-frequency partials.

Subtractive synthesis has been used to model percussion-like instruments and the human voice mechanism, but the results are somewhat inferior to those obtained with other source modellings techniques. The main limitation of subtractive synthesis is that non-linear interactions between the source and the resonator are ignored (i.e. there are no feedback interactions between the resonator and the source of excitation), thus imposing difficulties for controlling fine sound characteristics of acoustic instruments. For example, the frequency of the vibrating reed of a saxophone is strongly influenced by acoustic feedback from the resonating bore (i.e. 'tube') of the instrument, after being stimulated initially by a blast of air from the mouth. This phenomenon is almost impossible to simulate without feedback coupling. Nevertheless, subtractive synthesis is easy to understand, implement and control, and it can provide quite satisfactory results in some cases.

4.1.1 Brief introduction to filtering

In general, a filter is any device that performs some sort of transformation on the spectrum of a signal. For simplicity, however, in this section we refer only to filters which cut off or favour the resonance of specific components of the spectrum. In this case, there are four types of filters, namely: low-pass (LPF), high-pass (HPF), band-pass (BPF) and band-reject (BRF).

The basic building block of the subtractive synthesiser is the BPF, also known as the resonator. The BPF rejects both high and low frequencies with a passband in between. Two parameters are used to specify the characteristics of a BPF: passband centre frequency (represented as f_c) and resonance bandwidth (represented as bw). The bw parameter comprises the difference between the upper (represented as f_u) and lower (represented as f_l) cut-off frequencies (Figure 4.3).

The BRF amplitude response is the inverse of a BPF. It attenuates a single band of frequencies and discounts all others. Like a BPF, it is characterised by a central frequency and a bandwidth; but another important parameter is the amount of attenuation in the centre of the stopband.

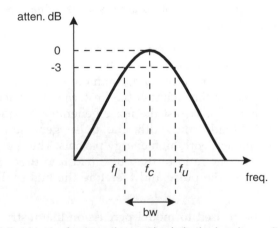

Figure 4.3 The basic building block of subtractive synthesis is the band-pass filter (BPF)

An LPF permits frequencies below the point called the cut-off frequency to pass with little change. However, it reduces the amplitude of spectral components above the cut-off frequency (Figure 4.4). Conversely, an HPF has a passband above the cut-off frequency where signals are passed, and a stopband below the cut-off frequency where the signals are attenuated. There is always a smooth transition between passband and stopband. It is often defined as the frequency at which the power transmitted by the filter drops to one half (about −3 dB) of the maximum power transmitted in the passband.

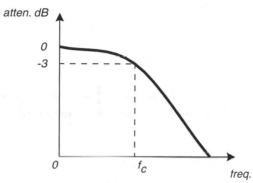

Figure 4.4 The low-pass filter (LPF) reduces the amplitude of spectral components above the cut-off frequency

Under special conditions a BPF may also be used as an LPF or an HPF. An LPF can be simulated by setting the BPF's centre frequency to zero. In this case, however, the cut-off frequency of the resulting low-pass is 70.7 per cent of the specified bandwidth, and not 50 per cent. For example, if the desired LPF cut-off frequency is to be 500 Hz, the bandwidth value for the BPF must be 1 kHz multiplied by 1.414 (that is, 1000 × 1.414). This is because the BPF is a two-pole filter; at its cut-off frequency (where output is 50 per cent) the output power of a true LPF of the same cut-off would be in fact 70.7 per cent.

An HPF can be made from a BPF by setting its centre frequency to be equal to the Nyquist frequency; i.e. the maximum frequency value produced by the system. HPFs made from BPFs suffer from the same approximation problems that affect LPFs.

The rate at which the attenuation increases is known as the slope of the filter – or *rolloff*. The rolloff is usually expressed as attenuation per unit interval, such as 6 dB per octave. In the stopband of an LPF with a 6 dB/octave slope, for example, every time the frequency doubles, the amount of attenuation increases by 6 dB. The slope of attenuation is determined by the order of the filter. *Order* is a mathematical measure of the complexity of a filter; in a digital filter, it is proportional to the number of calculations performed on each sample.

4.1.2 Filter composition

The resonator component of a subtractive synthesiser usually requires a composition of interconnected filters in order to produce the desired spectrum. There are two basic combinations for filter composition: *parallel connection* and *serial connection* (also known as *cascade connection*) (Figure 4.5).

In a serial connection, filters are connected like the links of a chain. The output of one filter feeds the input of the next, and so on. The output of the last filter is therefore the output of the entire chain. Much care must be taken when composing filters with different passband centre frequencies in series. In serial connection, the specification of one filter's passband does not guarantee that there will be significant energy transmitted in that passband. If any of the previous elements of the cascade has been significantly attenuated in that frequency range, then the response will be affected.

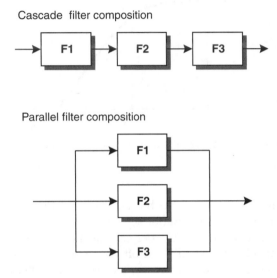

Figure 4.5 A subtractive synthesiser usually requires a combination of interconnected filters

With parallel connection, the signal to be filtered is simultaneously applied to all filter inputs. The outcome is the addition of the frequency responses of all filters, resulting in the resonation of any frequency found within the passband of the individual filters. Each filter of the parallel connection is usually preceded by an amplitude control, which determines the relative amplitudes of the partials of the resulting spectrum.

4.1.3 Subtractive synthesis examples

An in-depth study of the design of a subtractive formant (i.e. vocal) synthesiser in pcmusic is given in Chapter 6 and a few examples are available on the CD-ROM in folder *Vocsim*, in the pcmusic materials. Also, Koblo's Vibra 1000 for Macintosh (in folder *koblo*) features an analog-like synthesiser with simple but nevertheless efficient filtering capabilities.

4.2 Waveguide filtering

The waveguide filtering technique can be regarded as a filtering technique akin to subtractive synthesis, but unlike subtractive synthesis, waveguide filtering is based on the *feedback* interaction between a source of excitation and the filter component (Figure 4.6). This feedback interaction enables the waveguide technique to model resonating media other than acoustic chambers, such as strings. A waveguide filter comprises a variety of signal processing units, mostly delay lines and LPFs.

The best way to visualise the functioning of this technique is to imagine what happens when the string of a monochord is struck at a specific point: two waves travel in opposite directions and when they reach a bridge, some of their energy is absorbed and some is reflected back to the point of impact, causing resonance and interference (Figure 4.7).

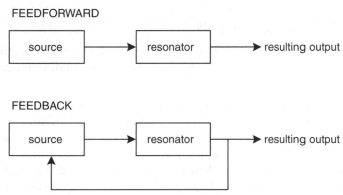

Figure 4.6 The feedback interaction between the source and the resonator is an important aspect of the waveguide technique

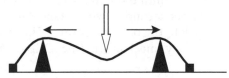

Figure 4.7 When the string of a monochord is struck at a specific point, two waves travel in opposite directions. When they reach a bridge, some of their energy is absorbed, and some is reflected back to the point of impact, causing resonance and interference

A generic waveguide filter instrument is illustrated in Figure 4.8. A source signal – usually noise – is input into a bi-directional delay line and travels until it reaches a filter (Filter A). This filter passes some of the signal's energy through and bounces some energy back; it models the effect of a 'scattering junction', such as a hole in a cylindrical tube or a finger pressing a string. The other filter (filter B) at the end of the chain models the output radiation type; for example, the bell of a clarinet.

The input source signal for the delay line and the type of modification applied by the waveguide chain play a key role in determining the characteristics of the instrument. Waves

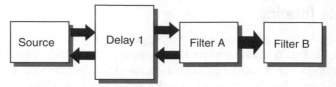

Figure 4.8 A generic waveguide filter instrument

running backwards and forwards along the chain cause resonance and interference at frequencies related to their dimensions. When the waveguide network is symmetrical in all directions, the sound it produces when stimulated will tend to be harmonic. If the waveguide twists, changes size, or intersects with another waveguide, then its resonant pattern is changed.

4.2.1 Waveguide filtering examples

Waveguide synthesis is largely compatible with most available sound synthesis programing languages; standard unit generators can often be used as the building blocks for waveguide networks.

Figure 4.9 illustrates a simple simulation of a pipe instrument implemented in Audio Architect; try *Pipe1.mae* in the folder *Aanets* in the Audio Architect materials on the accompanying CD-ROM. The model consists of a pipe open on one side and some kind of blowing mechanism on the other. When someone blows the instrument, a signal travels down the pipe. When this signal reaches the other end it is dissipated through the air creating disturbances that are perceived as sound. The laws of physics teach us that the air offers some resistance to disturbance thus making part of the outcoming energy bounce back inside the pipe towards the blowing source. The feedback portion then travels back towards the open end, and the loop continues until the signal fades away. Considering that the signal travels inside the pipe at one foot per second, it would take 8 milliseconds to reach the open end of the pipe and to bounce back to the blowing source. This phenomenon is simulated in *Pipe1.mae* by an 8-millisecond delay mechanism (the DELAY unit) that feeds back the delayed signal onto itself.

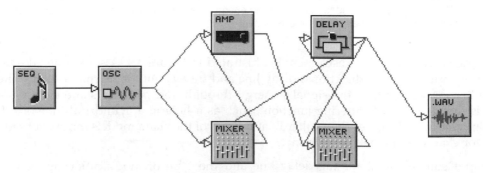

Figure 4.9 A simple simulation of a pipe instrument in Audio Architect

The blowing mechanism is simulated here by an oscillator (the OSC unit). We asssume that the lips of the musician send a periodic signal into the pipe and in order to create this signal the lips generate rapid streams of pressure bursts. When the signal bounces back, the lips could therefore be closed or open. If they are open the returning sound will collide with the newly produced burst thus producing a different reflection as if the lips were closed; the signals may even cancel each other. This is simulated by introducing an amplitude modulation arrangement (the AMP unit) before the signal is fed into the delay. Notice that the output from the oscillator plays a twofold role in *Pipe1.mae*: it acts as a source signal and as a modulator. The idea here is that the signal travels down the pipe and each time it bounces back it is amplitude modulated by the rapid opening and closing of the lips. By feeding the output of the delay back into itself and back to the modulator, one can roughly simulate the behaviour of the pipe and the action of the lips.

More sophisticated waveguide examples implemented in Nyquist are available on the CD-ROM in the folder *nyquist*; use a Web browser to open the file *example-home.htm* for explanations.

4.3 Karplus–Strong synthesis

Karplus–Strong synthesis uses a time-varying lookup table to simulate the behaviour of a vibrating medium. The basic functioning of this method starts with a lookup table of a fixed length, filled with random samples. In this case, the table functions as a queue of sample values, rather than as a fixed array, as in the case of a simple oscillator. As samples are output from the right side of the array they are processed according to a certain algorithm, and the result is fed back to the left side (Figure 4.10). The algorithm for processing the samples defines the nature of the simulation. For example, the averaging of the current output sample with the one preceding it in the array functions as a type of low-pass filter.

Although this 'recirculating wavetable' technique does not bear strong physical resemblance to the medium being modelled, its functioning does resemble the way in which sounds produced by most acoustic instruments evolve. They converge from a highly disorganised distribution of partials (characteristic of the initial noise components of the attack of a note) to oscillatory patterns (characteristic of the sustained part of a note).

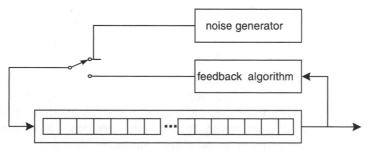

Figure 4.10 Karplus–Strong synthesis uses a time-varying lookup table to simulate the behaviour of a vibrating medium. As samples are output from the right side of the array, they are processed according to a certain algorithm and the results are fed back to the left side

The original Karplus–Strong algorithm, devised at Stanford University by Kevin Karplus and Alex Strong, averages the current output sample of a delay line with the preceding one, and feeds the result back to the end of the delay queue (Karplus and Strong, 1983). Note that in this case the delay line and the wavetable (or lookup table) are the same thing. This algorithm produces convincing simulations of the sound of a plucked string. The signal bursts with a loud, bright, percussive-like quality, then it darkens and turns into a simpler sinusoidal type of sound. The decay of the sound is normally dependent upon the size of the wavetable, but it is possible to control the decay length by adding time-stretching mechanisms to the original Karplus–Strong algorithm.

Indeed, several variations on the original Karplus–Strong algorithm have been developed in order to produce effects other than the plucked string. For example, simulations of drum-like timbres can be achieved by inverting the sign of a few output samples, according to a probability factor.

Some software packages for sound synthesis provide a special unit for the implementation of the Karplus–Strong technique, which normally needs five parameters to function:

- Amplitude
- Frequency
- Size of the recirculating lookup table
- The nature of the initialisation
- The type of feedback method

The size of the recirculating lookup table is specified in Hertz and is usually set to be equal to the frequency value. The lookup table can be initialised by either a random sequence of samples or a user-defined function. As for the feedback method, the simple averaging method described above is the most commonly used.

A few examples of Karplus–Strong synthesis in pcmusic are available in the folder *Pluck*, in the pcmusic materials on the CD-ROM. Check also for examples in Nyquist on the CD-ROM in the folder *nyquist*; use a Web browser to open the file *example-home.htm* for more information.

4.4 Cellular automata lookup table

Similarly to Karplus–Strong synthesis, the cellular automata lookup table also employs a recirculating wavetable to simulate the behaviour of a vibratory medium. The main difference is that the new sample values of the delay line are computed here by means of cellular automata, rather than by the averaging method.

Cellular automata are computer modelling techniques widely used to model systems in which space and time are discrete and quantities take on a finite set of discrete values (Wolfram, 1986). Physical systems containing many discrete elements with local interaction can be modelled using cellular automata. In principle, any physical system described by differential equations may be modelled as cellular automata.

Cellular automata are implemented as regular arrays of variables called cells. Each cell may assume values from a finite set of integers. Each value is normally associated with a colour.

The functioning of a cellular automaton is displayed on the computer screen as a sequence of changing patterns of tiny coloured cells, according to the tick of an imaginary clock, like an animated film. At each tick of the clock, the values of all cells change simultaneously according to a set of rules that takes into account the values of their neighbourhood (Figure 4.11). Although cellular automata normally have two or three dimensions, the recirculating lookup table technique uses only one-dimensional cellular automata, and the cells can assume only two values: on (represented by 1) or off (represented by 0).

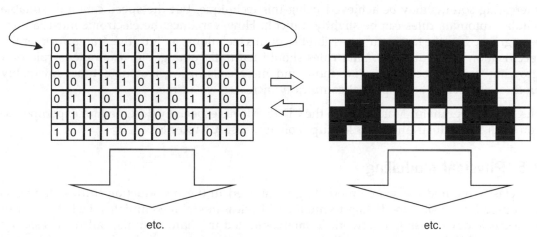

etc.   etc.

Figure 4.11 A cellular automaton is implemented as a regular array of variables called cells. Each cell may assume values from an infinite set of integers and each value is normally associated with a colour, in this case 0 = white and 1 = black

An example set of rules for the functioning of such types of automata can be defined as follows:

1 if 111 then 0
2 if 110 then 1
3 if 101 then 0
4 if 100 then 1
5 if 011 then 1
6 if 010 then 0
7 if 001 then 1
8 if 000 then 0

The eight possible states of the three adjacent cells (i.e. 3 adjacent cells = 2^3 rules) define the state of the central cell in the next step of the clock (Figure 4.12).

The synthesis method works by considering the array of cells of the one-dimensional cellular automaton as a lookup table; each cell of the array corresponds to a sample. The states of every cell are updated at the rate of the cellular automaton clock and these values are then heard by piping the array to the digital-to-analog converter (DAC).

Figure 4.12 Rule 3 states that if a cell is in state 0 and the state of its two neighbours is 1, then this cell remains in state 0 on the next tick of the CA clock

Interesting sounds may be achieved using this technique, but the specification of suitable cellular automata rules can be slightly difficult. However, once the electronic musician has been gripped by the fascinating possibilities of this technique, an intuitive framework for the specification of cellular automata rules should naturally emerge; for example, a rule of a 'growing' type of behaviour (i.e. more and more cells are activated in time) invariably produces sounds whose spectral complexity increases with time.

The LASy program for Macintosh on the CD-ROM was especially designed for experimentation with the cellular automata lookup table (see also Chapter 8).

4.5 Physical modelling

The classic method for source modelling mentioned in the introduction to this chapter is embodied by physical modelling techniques. Physical modelling emulates the behaviour of an acoustic device using a network of interconnected mechanical units, called *mass-spring-damping*, or MSD units (Figure 4.13). On a computer, this network is implemented as a set of differential equations, whose solution describes the waves produced by the model in operation. Sound samples result from the computation of these equations.

Figure 4.13 A string can be modelled as a series of masses connected by spring-damping units

This approach has the ability to capture two essential physical properties of vibrating media: *mass density* and *elasticity*. A single string, for example, can be modelled as a series of masses connected by spring-damping units (Figure 4.13). Consider that this string is in state of equilibrium. If force is applied to a certain point of the string (the equivalent to striking the string), then its masses are displaced from their initial positions. Disturbed mass units propagate the disturbance and generate wave motion by forcing adjacent masses to move away from their point of equilibrium (Figure 4.14). The compromise between the mass density and the elasticity of the model defines the speed of the propagation, the amount of the model's resistance to disturbance, and the time it takes to restore its equilibrium.

Surfaces (e.g. the skin of a drum) and volumes (e.g. the body of a guitar) can be modelled as a network of masses connected by more than one spring (Figure 4.15).

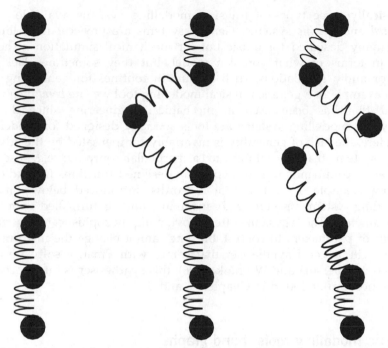

Figure 4.14 Disturbed mass units propagate the disturbance and generate wave motion by forcing adjacent masses to move away from their point of equilibrium (Note that this is only a schematic represention of the mass movements)

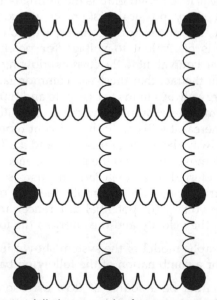

Figure 4.15 A membrane can be modelled as a matrix of masses connected by spring-damping units

There are basically three types of physical modelling systems available to musicians: *generic*, *dedicated* and *specific* systems. Generic systems are physical modelling tools that were not originally designed for music but for mechanical simulations. These tools are very flexible in terms of what you can model, but they sometimes lack the specific requirement for audio simulations, such as built-in routines for generating high-quality sounds. As an example of a generic physical modelling tool we cite *bond braphs*, a diagram-based method that has been used in mechanical engineering since the late 1950s. Dedicated physical modelling systems are tools specially designed for modelling musical instruments. Here, the lack of generality is invariably compensated by the advantages of a programming system that is tailored to a particular purpose; e.g. use of familiar terminology, fewer commands to learn, specially designed functions for audio, and so on. An outstanding example of such a tool is Cordis, introduced below. Finally, specific physical modelling systems provide ready-made instruments furnished with the means to control the parameters. These systems often provide highly sophisticated instruments with a great number of parameters to control, but one cannot change the instrument itself. As an example we cite a vocal synthesiser that comes with Praat, a software designed for phonetics research (Boersma and Weenink, 1996); this synthesiser is introduced below, and related information can be found in Chapters 6 and 8.

4.5.1 Generic modelling tools: bond graphs

Bond graphs are a diagram-based method, first introduced in the late 1950s, for representing physical systems (Paynter, 1961; Gawthrop and Smith, 1996). They consist of a set of standard elements connected together in a structure representative of the physical system being modelled. The connections between the elements are referred to as bonds, hence the name 'bond graphs'. On a basic level, bond graphs modelling is an extension and refinement of the premise that almost every physical system, be it acoustic, mechanic, hydraulic, thermodynamic or magnetic, has an electrical counterpart. Considering the example given earlier in Figure 4.1, force is equivalent to voltage (or potential difference) p over the capacitor C, and velocity is the equivalent to the current i flowing in the electric circuit. These equivalencies follow on from the fact that there is a common factor in all types of physical systems, namely *energy*. Energy always tends to dissipate and thus will *flow* anywhere where there is less energy at the time. For example, a warm room will not cool if all the adjacent rooms are equally warm, whereas it will cool if the adjacent rooms are cooler. When there is a flow of energy, there will always be a rate of this flow and a difference between the energy levels of the place the energy moves from and the place the energy moves to. In bond graphs parlance, this difference is referred to as the *effort*. In this case, the greater the effort, the greater the flow. In an electrical circuit, for example, the rate of flow is referred to as the current and the effort as the voltage (or potential difference); in a mechanical system, this rate would be referred to as the velocity and the effort as the force.

Figure 4.16 shows a bond graphs model of the system shown in Figure 4.1. A bond graphs model is composed of one or a combination of the following basic elements:

- **Energy source**: this can either be a source of flow or a source of effort; for example, pressure imposed on the mouthpiece of a clarinet, or gravity acting on a falling apple

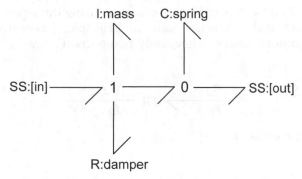

Figure 4.16 The corresponding bond graphs representation of the system shown in Figure 4.1

- **Energy store:** this can store either flow or effort. The accumulation of either flow or effort variables gives the system its state. The consequence is that the system's reaction to a certain input is dependent on the system's previous behaviour
- **Energy dissipater:** this is an element that simply dumps energy out of the system. For example, resistors and friction; both dissipate energy in the form of heat
- **Energy transfer:** these are elements that are responsible for transferring energy within the system. An energy transfer element can be either a *junction*, a *transformer* or a *gyrator*

There are two types of junctions: the *effort junction* and the *flow junction*. In an effort junction (or '0' junction), the effort on the bond coming to the junction is equal to the effort on each of the bonds coming out of the junction (Figure 4.17). In a flow junction (or '1' junction) the flow on the bond coming to the junction is equal to the flow on each of the bonds coming out of the junction (Figure 4.18).

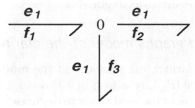

Figure 4.17 A common effort junction where $e_1 f_1 = e_1 f_2 + e_1 f_3$

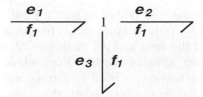

Figure 4.18 A common flow junction where $e_1 f_1 = e_2 f_1 + e_3 f_1$

Although Figures 4.17 and 4.18 portray only examples using three-port junctions, the reader should bear in mind that a junction can, in principle, have any number of ports. Transformers and gyrators, however, have only two ports (Figure 4.19).

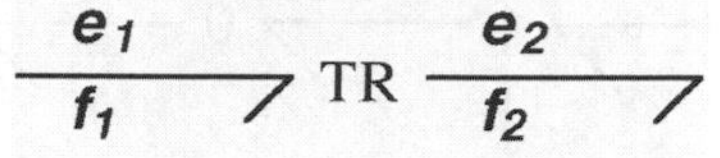

$$\frac{e_1}{f_1} \diagup \; TR \; \frac{e_2}{f_2} \diagup$$

Figure 4.19 The transformer element

The transformer (TR) specifies an output effort e_2 using the relationship $e_2 = ke_1$, where k is the transformer ratio. The value of f_2 is then forced to be $f_1 k$ so that $e_1 f_1 = e_2 f_2$. The gyrator (GY) is similar to the transformer, except that it specifies the output flow f_2 using the relationship $f_2 = ge_1$, where g is referred to as the mutual inductance.

Bond graphs are highly modular, lending themselves to the modelling of complex systems that can be decomposed into simpler subsystems at various levels. Within this context, the model of a musical instrument such as the clarinet can be specified by decomposing it into functional parts (e.g. mouthpiece, pipe, etc.), each of which is then modelled separately. The bond graphs that encapsulate each part of the instrument are then considered as a new bond graphs component that can be called on by any other high-level bond graphs structure (Figure 4.20). The bond graphs technique is appealing because it allows for the specification of models that visually resemble the structure of the mechanisms of the instruments they represent. Here, the system automatically produces the respective differential equations that govern the states of the model (i.e. the values for each energy store) and the output of the model.

The next section introduces an example of a bond graphs representation of a clarinet devised by Andrew McGregor, Peter Gawthrop and the author at the Department of Mechanical Engineering of Glasgow University (McGregor *et al.*, 1999).

An example study: A bond graphs model of the clarinet

When a musician blows the clarinet, the pressure in the mouth (P_o) becomes greater than the pressure in the mouthpiece (P_m), resulting in a flow of air through the aperture of the reed x (Figure 4.21). The aperture of the reed tends to decrease because the reed starts to close as soon as pressure is applied. While this is happening, a pressure wave is released into the pipe of the instrument. As the pressure wave travels down the pipe, the reed vibrates due to the stiffness of the reed, the effect of the flow of air U_f and the inertia. The effect of inertia dictates that the movement of the reed will overshoot a position of equilibrium, that is, the position where the two forces would balance out. The frequency of this initial vibration depends upon the stiffness of the reed and a function of U_f and when the pressure wave reaches the end of the pipe, they attempt to adjust themselves to the atmospheric pressure. The air on which the pressure wave is carried has mass and therefore inertia, when this equalisation takes place. Similar to the reed itself, this inertia causes the equalisation to overshoot and the result is that sections that were previously high pressure are now low pressure and vice versa; that is, the wave has been inverted. This inverted wave propagates itself from the end of the pipe of the instrument in all directions, including back up the pipe. The resulting superimposition of the original pressure waves travelling down the pipe and the inverted reflected waves travelling back to the pipe creates nodes (areas of alternation between high and low pressure) and antinodes (areas that stay at average pressure, typically atmospheric). A node cannot exist at the end of the clarinet bore or adjacent to a tone hole. Using this knowledge, it is possible to have control over the fundamental frequency (i.e. pitch) of the sound produced.

As the inverted reflected waves reach the mouthpiece of the clarinet, they cause the reed to open (when a high-pressure section of the wave reaches the reed) or close (when a low-pressure section of the wave reaches the reed). Since the resulting pressure wave emitted when the reed has been 'forcibly' opened or closed is larger than it would otherwise be, the period of the reed in motion becomes an exact divisor of the time it takes for a pressure wave to travel the length of the tube and back again. In this case, the pipe of the clarinet will tend to resonate only those partials that set up an antinode at the open end of the instrument; normally these are odd partials. It is this interaction between the pipe and the reed that forms the basis of the sound production of the clarinet.

Figure 4.20 The implementation of the clarinet model at its highest level

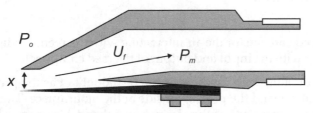

Figure 4.21 The mouthpiece of the clarinet

At the highest level, the clarinet model consists of a source-sensor (SS) component referred to as the player, that exerts an effort (i.e. pressure) on the reed component which is connected to the tube (or pipe) component. The tube imposes a pressure on a second SS component referred to as the listener. As the SS component is predefined by default (Gawthrop, 1995), the system already knows how it works, but the reed and the tube components need to be created.

The reed component

The reed component broadly follows the physical description given by de Bruin and van Walstijn (1995). However, rather than writing the model as a set of equations, it is graphically described by the bond graphs in Figure 4.22. The reed component consists of two sub-models:

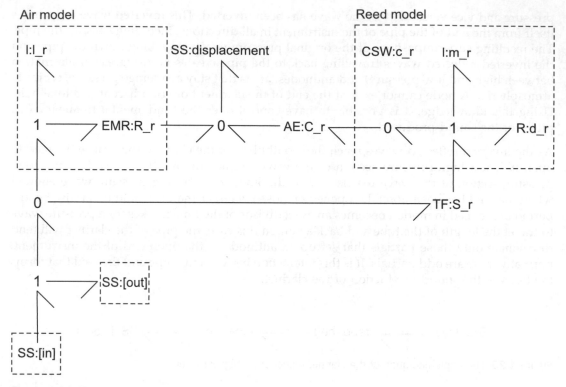

Figure 4.22 The reed component

one for the actual reed and one for the air interacting with the reed. As discussed below, these are coupled together with the input and output of the reed component.

The input and output of the reed component can be thought of as the pressure applied to the mouthpiece by the player and the pressure inside of the mouthpiece. These correspond to the two ports of the reed component of the clarinet model and are specified as SS components. As the flow of air will be the same for both SS components, they are linked by a common flow function, that is, a '1' junction, in bond graphs parlance. As explained earlier, the effort carried by the bond entering the junction will equal the sum of the effort carried by the bonds exiting the junction. This means that it is the difference in pressure inside and outside the mouthpiece that drives the movement of the reed.

The reed is modelled as a classic MSD structure: the mass, spring and damping components are labelled m_r, c_r and d_r, respectively, and they share the common velocity of the '1' junction with that of the air immediately adjacent to the reed associated with the bond attached to the TF component labelled S_r. This TF component transforms the volumetric flow of air U_f associated with the reed motion to the corresponding velocity x (x is a time derivative).

This '1' junction and the TF component ensure the correct kinematic relationship between the MSD components and the additional volume flow due to reed motion. Moreover, and this is a key attribute of the bond graphs technique, the fact that the '1' junction and the TF are energy-conserving automatically gives the correct relationship between the corresponding force and pressure. This is a great help for obtaining correct models. Similarly, the '0' junction at the other side of the TF component ensures the correct pressure and flow relationship between the reed and the mouthpiece.

The air sub-model has two components: the inertia of the air above the reed and the (non-linear) air flow resistance labelled I_r and R_r, respectively. The former is the straightforward bond graphs I component (mass) considered earlier, but the other component is more complex, as the flow resistance is modulated by the reed position. Thus R_r is not the simple R component (dumper) but rather an Effort-Modulated Resistor (EMR) component (Figure 4.23) which is modulated by the displacement of x of the reed; please refer to book by Gawthrop and Smith (1996) for more information on bond graphs components. The fact that displacement appears as an effort in this context is a result of the way we have chosen to model the system. The constitutive relationship embedded in this EMR component follows de Bruin and van Walstijn's description (1995).

At this point, it is important to include a significant feature of the model: the *reed displacement* $x > x\_0$ (x_0 is a constant) as its motion is constrained by the mouthpiece. This is achieved by using a Switched C (CSW) component in place of the standard bond graphs C component (spring). The CSW component (Figure 4.24) is an extension of the standard bond graphs C component that allows for discontinuous (or 'switching') behaviour (Gawthrop, 1995). In this context, it is equivalent to a constrained spring which is not allowed to move beyond a certain point.

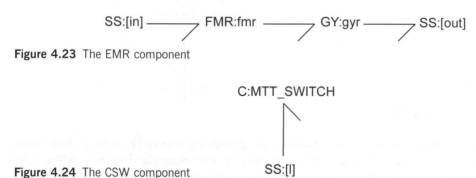

Figure 4.23 The EMR component

Figure 4.24 The CSW component

In this model, it is assumed that the reed is driven only by the air pressure, rather than by the air velocity. This appears in the bond graphs model via the effort amplifier (AE) component which prevents direct air velocity/reed interaction. A more complex model could indeed include such an interaction, but that is beyond the scope of this example.

The tube component

The tube component is a much simpler model than the reed. As a cross-section of air in a pipe has an elasticity and mass, and it loses energy in the form of friction when it moves, the model simply consists of a series of MSDs and TFs (Figure 4.25).

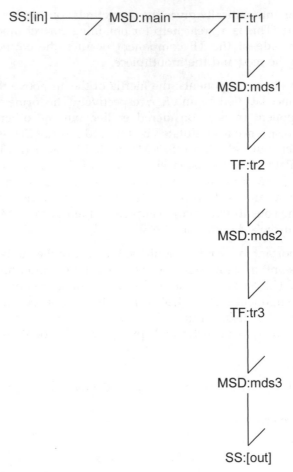

Figure 4.25 The tube component

This can be implemented as an approximation of the partial differential equations describing the motion of the air column. For the sake of simplicity, this example does not include the effect of the air holes dynamically opening and closing; this could be done simply by adding the appropriate bond graphs components to the model.

Bond graphs models are capable of producing realistic emulations of acoustic instruments. They have been designed as a general tool for physical modelling, not necessarily limited to musical instruments (e.g. aircraft aerodynamics). However, this generality may not be always necessary or even desired. In most cases, a tool that is focused on the task at hand can be more productive and easier to grasp.

A number of systems exist for bond graphs modelling. Our clarinet example was implemented using MTT (Model Transformation Tools), a tool developed by Professor Peter Gawthrop at the University of Glasgow. MTT was designed for modelling dynamic physical systems using bond graphs and transforming these models into representations suitable for analysis, control and simulation. MTT works in combination with Octave, a high-level language, primarily intended for solving linear and non-linear problems. Octave has extensive tools for solving common numerical linear algebra problems, finding the roots of non-linear equations, integrating ordinary functions, manipulating polynomials, and integrating ordinary differential and differential-algebraic equations. It is easily extensible and customisable via user-defined functions written in Octave's own language, or using dynamically loaded modules written in C++, C, Fortran, or other languages. Neither MTT nor Octave is provided on the CD-ROM, but those readers interested in diving into serious physical modelling are welcome to consult Professor Gawthrop's Web site for more information. His Web address plus some links to Web sites dedicated to bond graphs are available on the accompanying CD-ROM; use a Web browser to open the *web-refs.html* file, in folder *various*.

4.5.2 Dedicated modelling systems: Cordis and Genesis

Genesis is a physical modelling system dedicated to building musical instruments, developed by a team of researchers led by Claude Cadoz at ACROE (Association pour la Création dans les Outils d'Expression) in Grenoble, France (Cadoz et al., 1984; Cadoz and Florens, 1990, 1991). In fact, Genesis is a visual interface for Cordis-Anima, a programming language that ACROE developed for implementing simulations of sounds (Cordis) and animated objects (Anima). The following paragraphs present some fundamental notions of Cordis, followed by an introduction to Genesis.

Essentially, Cordis instruments are built using two types of 'atoms': *matter* and *links* (Figure 4.26). For the sake of simplicity, we will not delve into the inner details of how these atoms are implemented; it suffices to assume that an atom is the smallest unit of the Cordis universe.

Matter is considered to exist in three-dimensional space and its state is computed according to forces exerted by the surrounding environment. A link, as its name suggests, is used to connect pieces of matter: it acts as a mechanical physical bridge between two pieces of matter. Note that a link does not actually take up physical space.

Matter and link atoms are effectively connected by two-channel ports: one channel is used to communicate force vectors, represented as *F*, and the other to communicate displacement

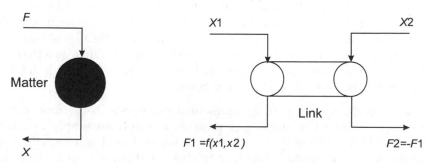

Figure 4.26 Matter and links are the two atoms of the Cordis universe

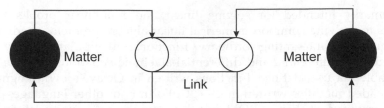

Figure 4.27 A Cordis instrument consists of a network of pieces of matter and links

vectors (or positions) represented as X. Cordis instruments therefore consist of a network of pieces of matter bridged by links (Figure 4.27).

There are three types of links in Cordis: *spring, friction* and *conditional*. Each of these implements a specific type of bridge between matter. Spring and friction implement the standard spring and damping elements mentioned earlier. The conditional link implements a combination of spring and friction, and it allows for the specification of non-linear interactions in the network.

As a graphic interface for Cordis, Genesis presents itself to the user with a 'canvas' and a menu of atoms. A piece of matter is represented by a circle and a link is represented by straight wire. A network is built by dragging the respective atoms from the menu onto the 'canvas' and connecting them. Each atom has a number of attributes that can be edited via a pop-up menu.

Genesis has two types of atoms on the menu: MAT (matter) and LIA ('liaison', or link in English). Basically, there are two kinds of MAT: MAS and SOL. Matter is represented by coloured circles; by default MAS is orange and SOL is green (the user can customise these colours). MAS stands for mass and it has three parameters: inertia M, initial position $X0$ and speed $V0$. A mass will preserve its position and speed if no force is applied to it. SOL is an element that cannot move because its mass is infinite and its velocity is always zero (that is, its position never varies). SOL (or 'ground' in English) is normally used to represent a fixed support for an object.

As for LIA elements, there are five types: RES ('ressort', or spring in English), FRO ('frottement', or friction in English, also known as 'damping'), REF ('ressort-frottement', or spring-friction in English, generally referred to as spring-dumping), BUT ('butée', or block in English) and LNL ('liaison non-linéaire', or non-linear link in English). LIA elements define the nature of the bridge between two MATs and each type has a set of coefficients that needs to be specified. For instance, REF (a combination of RES and FRO), has two coefficients: stiffness (of the spring) K and force absorption (of the damper) Z. Whereas a BUT establishes that one MAS will strike another MAS and will bounce away, a LNL establishes more complex strike methods, such as plucking and bowing.

Figure 4.28 shows an example of a string represented in Genesis. This string could be set to vibrate by pulling out one of its masses. In practice, this is done simply by changing the initial position $X0$ of one MAS of the string. A more sophisticated way to set the string into vibration would be to implement a hammer to strike it (Figure 4.29). A hammer can be implemented by a MAS bridged to a MAS of the string by means of a BUT.

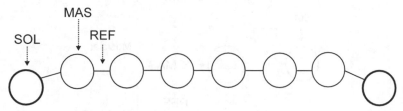

Figure 4.28 An example of a string represented in Genesis

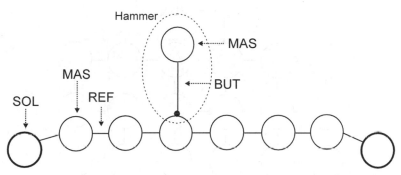

Figure 4.29 An example of an instrument consisting of a string and a hammer to strike it

The example of the hammer in Figure 4.29 hints at a very interesting aspect of Genesis: the ability to build hierarchical vibrating structures whereby the behaviour of the one can control the behaviour of another. In the case of Figure 4.29, the string could be considered as a vibrating structure which is activated by a higher-level device, the hammer. As far as Genesis is concerned, both string and hammer are vibrating structures, but the oscillation periods are of different scales. Indeed, one could build an even higher level device here to drive the hammer, and so forth. Figure 4.30 shows a three-level structure: a membrane (implemented as an array of masses), a hammer (implemented in the form of a simple oscillator bridged to the membrane by means of a BUT link) and a 'musician' (a device that controls the hammer). Genesis allows for the specification of complex structures involving more than one vibrating body, various excitations and musicians, and a conductor, all in one single 'instrument'. What is interesting here is that Genesis encourages composers to think in terms of a unified physical modelling framework for composing musical structures. Claude Cadoz has recently composed a piece called *pico . . . TERA*, which is a 290-second piece entirely generated by one run of one such complex instrument.

Cordis and Genesis currently run only on Silicon Graphics computers under the Irix operational systems. The ACROE team is producing versions for Linux and Mac OS X; check the current status of this on their Web site, whose address can be found on the CD-ROM in the *web-refs.html* file, in the folder *various*. A number of diagrams and sound examples, including an exclusive excerpt of Cadoz's pioneering piece *pico . . . TERA*, can be found in the folder *genesis*.

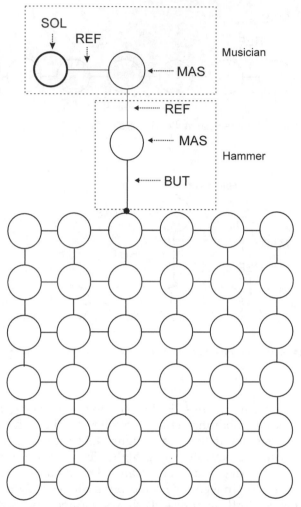

Figure 4.30 A three-level structure instrument consisting of a membrane, a hammer and a device that controls the hammer, or 'musician'

4.5.3 Specific physical model implementations: Praat's articulatory synthesiser

Praat, a system originally designed for phonetics research at the University of Amsterdam, comes with a remarkable physical model of the vocal tract, referred to as the *articulatory synthesiser*. The synthesiser is fully programmable, but we cannot change its architecture; we can only write scores or tweak its parameters via a graphic user interface (GUI). A complete version of Praat is available on the CD-ROM. The functioning of the vocal tract is discussed below, followed by an introduction to the architecture of the synthesiser; please also refer to Chapters 6 and 8 for more details.

Understanding the vocal tract

The vocal system can be thought of as a resonating structure in the form of a complex pipe (Figure 4.31). The acoustic properties of a straight regular pipe are well understood and fairly straightforward to model, but the shape of the vocal tract is neither straight nor regular. On the contrary, it is bent, its length can be stretched and contracted, its diameter is variable at different sections and it branches into two main paths: one to the mouth and the other to the nasal cavity. Moreover, length, diameter and branching control (for producing nasals) change during sound production. These properties challenge the design of a realistic physical model of our vocal mechanism.

Another aspect that increases the challenge of the modelling task is that there are a number of variable obstructions inside the pipe, such as the vocal folds and the tongue. These give the vocal pipe discontinuities whose acoustic reflections are cumbersome to simulate. Also, one should not dismiss the importance of the elastic properties of the walls (and the internal obstructions of the pipe) for the acoustic behaviour of the vocal system, in the sense that these elements vibrate sympathetically during sound production. The nature of these vibrations obviously depends upon the elasticity of the vibratring media but the difficulty here is that elasticity differs according to specific circumstances during sound production.

For modelling purposes, it is acceptable to think of the vocal tract as a resonating pipe system, as long as the model takes into account that:

- The form of the pipe is variable
- The pipe contains changeable internal obstructions
- The elastic properties of the walls of the pipe and of its internal obstructions can alter during sound production

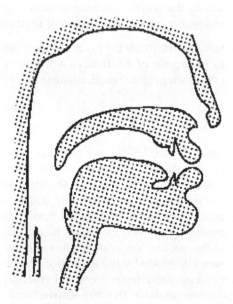

Figure 4.31 For modelling purposes the vocal tract can be regarded as a resonating pipe system

A plausible approach to modelling the vocal system is to consider it as a structure of chained short pipes, each representing a different section of the compound resonating system. The model includes a closed pipe, representing the bottom end at the lung level, and a number of open pipes, which are either chained to other pipes, or are open to the atmosphere. Two types of connection would be possible for building pipe structures using these resonating building blocks: branched and non-branched. In the former case, the open boundary of a pipe acts as an interface with two other pipes. This happens, for example, when the pharynx branches into the oral and nasal cavities at the velopharyngeal port. Non-branched connection involves only two pipes chained one after the other.

Traditionally, the vocal system has been modelled as a subtractive synthesiser consisting of a source module and a resonator module (see the example given in section 6.2, Chapter 6). Basically, the source module produces a raw signal intended to simulate the waveform produced by the vibration of the vocal folds, which in turn is shaped by the acoustic response of the resonator module. The source is normally implemented using pulse and white noise generators, plus a low-pass filter arrangement. The result is a signal that resembles the waveform produced by the vibration of the vocal folds. The resonator is implemented as an arrangement of band-pass filters, each of which is tuned to resonate at the frequencies of the most prominent standing waves of the irregular pipe. These standing waves correspond to the distinct resonance – or formant configuration – that confers the characteristic timbre of the human voice. In this case, each filter, or group of filters, would roughly correspond to a section of the compound vocal pipe (Klatt, 1980).

Subtractive formant synthesis is fine for producing vowels but it is not so effective for consonants. It is inadequate for producing realistic concatenations of syllables, not to mention paralinguistic utterances such as the falsetto tone of voice, and the trills and clicks used in various African languages (Laver, 1994). The limitations are even more restrictive in musical applications, especially in the realm of contemporary music where composers often seek to extrapolate the sonorities of conventional musical instruments.

There have been several attempts to minimise the limitations of subtractive synthesis, ranging from better filtering strategies to the use of alternative source modules. However successful these attempts have been, none of them matches the realism of a physical model.

Praat's physical model

Boersma (1991), Cook (1992) and Maeda (1982) have reported significant progress in physical modelling of the vocal tract. Maeda modelled the vocal tract as a straight single pipe approximately 17 cm long, with variable circular cross-sections. Waves less than about 4 kHz propagate in this model, resulting in a digital equivalent of a classical analog transmission-line representation of a resonating system; see Olson's book *Musical Engineering* for more details on such analog representations (1952). Cook has modelled the vocal tract as a waveguide network; refer to the section on waveguides in this chapter. Both Cook's and Maeda's architectures, however, still abide by the excitation/filter paradigm in the sense that a voicing excitation is posited separately from the vocal tract representation. This imposes limitations when it comes to simulating the myoelastic and aerodynamic interactions involving the larynx and the resonating vocal tract.

Boersma, one of the authors of Praat, modelled the whole vocal system as a pipe composed of dozens of smaller pipes whose walls were modelled using MSD units; each of these small pipes has four walls. The model makes no distinctions between source and resonating modules in the sense that the whole vocal system, including the larynx and the vocal folds, is represented as a composite pipe whose components (or 'sub-pipes') are modelled using identical MSD units (Figure 4.32). The leftmost of the sub-pipes is closed (as if the closed boundary were the diaphragm), and the rightmost sub-pipes (corresponding to the mouth and the nostrils) are open to the atmosphere. Roughly speaking, the springs in Figure 4.32 represent muscles in the sense that their rest positions and tensions can be altered in order to adjust the walls and the internal obstructions of the pipe. The central idea here is that the walls and obstructions yield to air pressure changes and air is forced to flow inside the pipe as the result of mass inertia and elasticity. For example, by diminishing the volume of the lungs, one generates displacement of air towards the mouth; in other words, excitation for phonation.

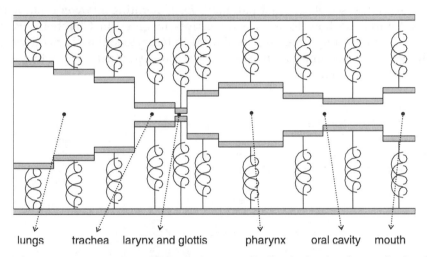

lungs	trachea	larynx and glottis	pharynx	oral cavity	mouth

Figure 4.32 Schematic representation of the vocal system. Each sub-pipe has four walls, but for the sake of clarity, the third dimension is not represented in this figure. Also omitted here is the branching of the nose at the boundary between the pharynx and oral cavity

Figure 4.33 depicts the cross-section of a single sub-pipe. The damping element is not shown in Figure 4.32, but each spring of the model does have a damping element associated with it. The articulatory muscles of the model control the rest position of the wall (by changing the values of the spring constant and damping) and the tube length.

Two types of equation are needed to compute the network: *myoelastic equations* that describe the physical behaviour of the walls of the pipes and *aerodynamic equations* that describe the evolution of the movements and pressures of the air in the network. The variables for these equations are specified by means of 29 parameters that metaphorically describe the actions of the vocal tract muscles and organs, such as cricothyroid, styloglossus, orbicularis oris and masseter, to cite but four. This synthesiser is capable of producing utterances with an

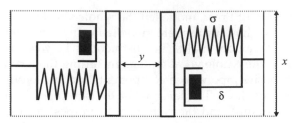

Figure 4.33 Cross-section of a single sub-pipe

incredible degree of realism. However, there is a price to pay for such sophistication: it is prohibitively tedious and time-consuming to specify the right movements of the muscles for producing required utterances. Also, it is computationally very demanding. A 5-second long sound can take as long as 15 minutes to compile on a modest Pentium MMX processor. An example of a score (or 'script' in Praat parlance), for generating a vowel is shown below; an introductory tutorial on how to write such scripts in given in Chapter 6. Basically this example sets values for lung pressure, vocal folds (interarytenoid and cricothyroid), tongue position (hyoglossus) and mouth (masseter). The non-specified parameters are set with default values by the system.

```
# --------------------------------
# Computer Sound Design
# E. R. Miranda - Focal Press
# --------------------------------
Create Speaker . . . morag Female 2
Create Artword . . . vowel-a 0.5
# ================================
# SUB-GLOTTAL AND GLOTTAL CONTROL
# ================================
# -----------------------
# Lung pressure
# -----------------------
Set target . . . 0.0   0.12  Lungs
Set target . . . 0.05  0.0   Lungs
Set target . . . 1.0   0.0   Lungs
# ------------------------------------------------------
# Glottis closure 0.5 = phonation 1.0 = stiff 0.0 = relaxed
# ------------------------------------------------------
Set target . . . 0.0  0.5  Interarytenoid
Set target . . . 1.0  0.5  Interarytenoid
# -----------------------
# Glottis: 1 = higher pitch
# -----------------------
Set target . . . 0.0  0.5  Cricothyroid
Set target . . . 1.0  0.5  Cricothyroid
```

```
# ==========
# VOCAL TRACT
# ==========
# For all muscles 0.0 = Relaxed/Neutral position
# -----------------------------------
# Hyoglossus = Tongue back and downwards
# -----------------------------------
Set target . . . 0.0  0.45  Hyoglossus
Set target . . . 1.0  0.45  Hyoglossus
# ==========
# MOUTH
# ==========
# ------------------------------------------------
# Masseter = Jaw Opening -0.5 = open; 0.5 = closed
# ------------------------------------------------
Set target . . . 0.0  -0.45  Masseter
Set target . . . 1.0  -0.45  Masseter
#
# === Synthesise ===
#
select Speaker morag
plus Artword vowel-a
To Sound . . . 44100 25  0 0 0  0 0 0  0 0 0
```

Praat exists for many platforms, Linux, Unix, Windows and Mac OS. Fully working versions for Windows and Mac OS are available on the CD-ROM in folder *praat*. An impressive user manual, accessed via the program's Help menu, is also provided.

4.6 Modal synthesis

Modal synthesis is similar to physical modelling in many respects and the literature often regards modal synthesis as a particular technique for implementing physical models. It is not our obejctive here to foster a discussion about what should or should not be regarded as physical modelling. From a practical standpoint, it would suffice to consider that where modal synthesis differentiates from physical modelling is that the level of abstraction of the former is higher than the level of abstraction of the latter. The level of abstraction of physical modelling occurs at the MSD units. For example, in a system such as Genesis, one builds instruments with low-level MAT and LIA components. The level of modal synthesis is high in the sense that here one builds an instrument by combining pre-fabricated components, or *modal acoustic structures*. In modal synthesis, instruments are built with components such as membranes, tubes, wires and bridges (Figure 4.34).

In fact, one could encapsulate physical modelling networks into modules and provide them in a library. But in physical modelling we have access to the units that make up these modules, whereas in modal synthesis we cannot change their inner structure.

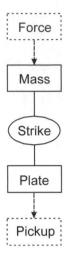

Figure 4.34 The level of abstraction of modal synthesis is higher that the level of abstraction of physical modelling in the sense that the building blocks of modal synthesis are prefabricated vibratory structures

From a technical point of view, it is what is inside the prefabricated components that makes modal synthesis distinct from physical modelling. In simple terms, a modal component can be thought of as a 'black box' that embodies certain vibratory properties. These vibratory properties are characterised in terms of *modal information*. Modal information describes the modes of vibration of an acoustic system. In order to visualise this, imagine that modes are associated with the harmonic response of an open pipe. In this case, the standing waves of the pipe correspond to the modes of the pipe, where the lowest frequency standing wave is the first order mode and its multiples are higher order modes. The representation of a modal synthesis component encodes the shape of the respective vibrating structure, its modal frequencies and damping coefficients. Thomas Rossing's book *The Science of Sound* presents a good introduction to modes in vibrating systems (Rossing, 1990).

The Mosaic system, developed by Jean-Marie Adrien at the University of Paris VI, is a typical example of a tool for implementing modal synthesis (Adrien, 1991). It provides a number of ready-made substructures such as strings, air columns, metal plates and membranes, as well as substructures for the simulation of actions such as bowing and hammering. Instruments are thus programmed by networking these substructures. In Mosaic terms, substructures are called *objects* and the interactions between objects are referred to as *connections*. A connection between two objects also acts as an interface between the user and the instrument; for example, the connection between a bow and string provides the means for user-customisation of the parameters that drive the interaction, such as pressure, speed, etc.

It is always hard to compare synthesis methods without first establishing what needs to be achieved; what may work satisfactorily for one case might not for another. In comparison to physical modelling, the modal approach has the advantage of the reduction of mathematical complexity and the modularity of the substructures. The instrument designer can add or subtract substructures on a network to create time-varying synthesis effects (such as expanding or shrinking the size of an instrument) and/or timbral hybrids by combining substructures from different instruments. But again, if the composer wishes to work with MSD structures and/or explore the physical modelling ideal to generate musical form, then the Genesis system might prove a better option than Mosaic.

Mosaic has been further developed by Jean-Marie Adrien himself and a number of other people at Ircam, and it is now available for Unix and Mac OS platforms under the name Modalys. Modalys instruments are built using an interface language called Scheme, whose syntax is very similar to Lisp. There is also Modalys-ER, a GUI for programming Modalys. For more information about Modalys please consult Ircam's Web site available in *web-refs.htm*, in the folder *various* on the accompanying CD-ROM.

5 Time-based approaches: from granular and pulsar to PSOLA and statistical

Philosophical considerations aside, time is one of the most important parameters of sound. The great majority of acoustic, synthesis and music descriptors depend upon some timing factor; e.g. wavecycle, frequency, samples, sampling rate, rhythm, etc. It is therefore not surprising that some synthesis techniques approach the problem from a time-domain perspective.

In a lecture entitled *Four Criteria of Electronic Music*, given in London in 1971, the composer Karlheinz Stockhausen introduced some inspiring time-domain sound-processing methods that played an important role in the development of computer tools for time modelling (Maconie, 1989). The general idea of Stockhausen was that if two different pieces of music, such as a symphonic piece by Villa Lobos and a Scottish folk tune, are accelerated until they last only a couple of seconds each, the results would be two different sounds with very distinct timbres. Conversely, if two different short sounds are stretched out in time, then the result would be two different sound streams whose forms reveal the inner time structures of the original sounds. Based upon this idea, in the 1950s Stockhausen devised a very peculiar way to synthesise sounds, which is perhaps the first case of a time-modelling technique. He recorded individual pulses on tape, from an electronic generator. Then he cut the tape and spliced the parts together so that the pulses could form a particular rhythm. Next he made a tape loop of this rhythm and increased the speed until he could hear a tone. Various tones could be produced by varying the speed of the loop. Different rhythms on tape produced different timbres; the components of the rhythmic sequence determined the spectrum according to their individual cycle on the tape loop.

The most interesting aspect of this method, from a composer's point of view, is that it encourages one to work with rhythm and pitch within a unified time domain. It should be noted, however, that the transition from rhythm to pitch is not perceived precisely. The human ear can hear a distinct rhythm up to approximately 10 cycles per second but a distinct pitch does not emerge until approximately 16 cycles per second. It is therefore not by chance that the categorical differentiation between rhythm and pitch has firmly remained throughout the history of Western music.

On the whole, Stockhausen's inventive work gave rise to a number of approaches to current time-modelling synthesis. The postulate here is that sounds may be specified entirely by the variations of cyclic patterns in the time domain. In this case, musically interesting sounds may be synthesised by forging the transformations of its waveform as they develop in time. In a broader sense, time-modelling synthesis thus uses techniques that control wave transformations. Time modelling is particularly suited for the synthesis of highly dynamic sounds in which most of its properties vary during their course.

5.1 Granular synthesis

Granular synthesis works by generating a rapid succession of very short sound bursts called *grains* (e.g. 35 milliseconds long) that together form larger sound events. The results generally exhibit a great sense of movement and sound flow. This synthesis technique can be metaphorically compared with the functioning of motion pictures in which an impression of continuous movement is produced by displaying a sequence of slightly different images at a rate above the scanning capability of the eye.

The composer Iannis Xenakis is commonly cited as one of the mentors of granular synthesis. In the 1950s, Xenakis developed important theoretical writings where he laid down the principles of the technique (Xenakis, 1971). The first computer-based granular synthesis system did not appear, however, until Curtis Roads (1988) and Barry Truax (1988) began to investigate the potentials of the technique systematically in the 1970s, in the USA and Canada respectively.

5.1.1 The concept of sound grains

In order to gain a better understanding of granular synthesis it is necessary to comprehend the concept of *sound grains* and their relationship to our auditory capacity. This notion is largely based upon a sound representation method published in 1947, in a paper by the famous British physicist Dennis Gabor (1947). Until then the representation of the inner structure of sounds was chiefly based upon Fourier's analysis technique and Helmholtz's musical acoustics (Helmholtz, 1885).

In the eighteenth century, Fourier proposed that complex vibrations could be analysed as a set of parallel sinusoidal frequencies, separated by a fixed integer ratio; for example, $1x$, $2x$, $3x$, etc., where x is the lowest frequency of the set. Helmholtz then developed Fourier's analysis for the realm of musical acoustics. He proposed that our auditory system identifies musical timbres by decomposing their spectrum into sinusoidal components, called *harmonic series*. In other words, Helmholtz proposed that the ear naturally performs Fourier's analysis to distinguish between different sounds. The differences are perceived because the loudness of the individual components of the harmonic series differs from timbre to timbre. The representation of such analysis is called *frequency-domain representation* (Figure 5.1). The main limitation of this representation is that it is timeless. That is, it represents the spectral components as infinite, steady sinusoids.

Although it is taken for granted now that the components of a sound spectrum vary substantially during the sound emission, these variations were not considered relevant enough to be represented until recently. In his paper, Gabor proposed the basis for a representation method which combines frequency-domain and time-domain information. His point of departure was to acknowledge the fact that the ear has a time threshold for discerning

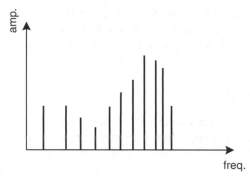

Figure 5.1 Frequency-domain representation is timeless because it represents the spectrum of a sound as infinite, steady sinusoids

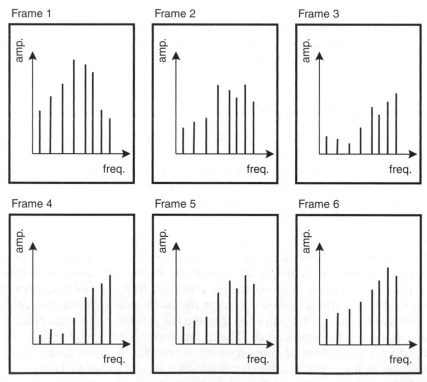

Figure 5.2 Frequency-domain representation snapshots can show how the spectrum of a sound evolves with time

sound properties. Below this threshold, different sounds are heard as clicks, no matter how different their spectra might be. The length and shape of a wavecycle define frequency and spectrum properties, but the ear needs several cycles to discern these properties. Gabor called this minimum sound length *acoustic quanta* and estimated that it usually falls between 10 and 30 milliseconds, according to the nature of both the sound and the subject.

Gabor's theory fundamentally suggested that a more accurate sound analysis could be obtained by 'slicing' the sound into very short segments, and by applying a Fourier type of analysis to each of these segments. A sequence of frequency-domain representation snapshots, one for each segment, would then show – like the frames of a film – how the spectrum evolves with time (Figure 5.2). Granular synthesis works by reversing this process; that is, it generates sequences of carefully controlled short sound segments, or *grains*.

5.1.2 Approaches to granular synthesis

As far as the idea of sound grains is concerned, any synthesiser producing rapid sequences of short sounds may be considered as a granular synthesiser. However, there are important issues to consider when designing a granular synthesis instrument. Three general approaches to the technique are presented as follows.

Sequential approach

The sequential approach to granular synthesis works by synthesising sequential grain streams. The length of the grains and the intervals between them are controllable, but the grains must not overlap (Figure 5.3).

time

Figure 5.3 The sequential approach to granular synthesis works by synthesising sequential grain streams

Scattering approach

The scattering approach uses more than one generator simultaneously to scatter a fair amount of grains, not necessarily in synchrony, as if they were the 'dots' of a 'sonic spray'. The expression *sound clouds* is usually employed by composers to describe the outcome of the scattering approach (Figure 5.4).

Granular sampling approach

Granular sampling employs a *granulator* mechanism that extracts small portions of a sampled sound and applies an envelope to them. The granulator may produce the grains in

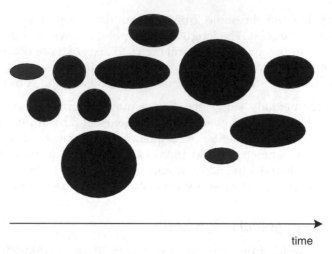

time

Figure 5.4 The scattering approach to granular synthesis uses more than one generator simultaneously to scatter a fair amount of grains, not necessarily in synchrony

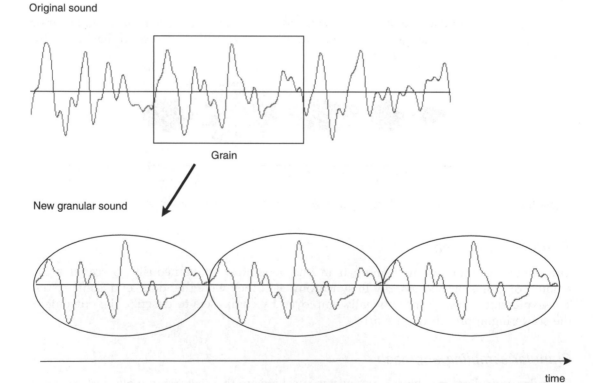

Original sound

Grain

New granular sound

time

Figure 5.5 Granular sampling works by replicating a small portion extracted from a sampled sound

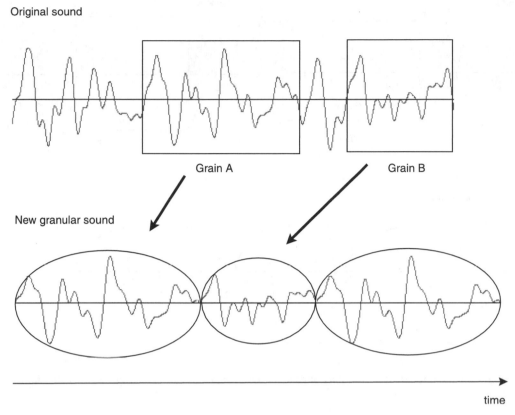

Figure 5.6 A granular sampler may also extract grains from different parts of the sample

a number of ways. The simplest method is to extract only a single grain and replicate it many times (Figure 5.5). More complex methods involve the extraction of grains from various portions of the sample. In this case, the position of the extraction can be either randomly defined or controlled by an algorithm (Figure 5.6). Interesting results can be obtained by extracting grains from more than one sound source (Figure 5.7). Other common operations on 'granulated' sounds include: reversing the order of the grains, time expansion (by changing the time interval between the grains) and time compression (by truncating grains). On the accompanying CD-ROM (in the folder *crusher*) there is crusherX-Live, a granular sampling system for PC-compatible under Windows platforms.

5.1.3 Designing granular synthesisers

In granular synthesis, sound complexity is achieved by the combination of large amounts of simple grains. The architecture of a granular instrument may therefore be as simple as a single oscillator and a Gaussian-like bell-shaped envelope (Figure 5.8).

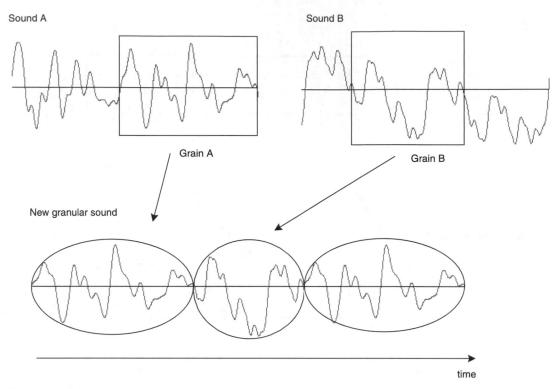

Sound A

Sound B

Grain A

Grain B

New granular sound

time

Figure 5.7 Interesting granular sampling effects can be obtained by combining grains extracted from more than one sound source

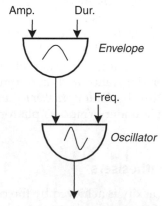

Amp.   Dur.

Envelope

Freq.

Oscillator

Figure 5.8: A granular synthesis instrument may be as simple as a single oscillator and a Gaussian-like bell-shaped envelope

The envelope is of critical importance in granular synthesis because it prevents the glitches that would otherwise be produced by possible phase discontinuities between the grains (Figure 5.9). The shape of the envelope should be carefully chosen because unwanted distortions may be produced by unavoidable 'envelope modulation' (i.e. ring modulation) effects (see Chapter 2). Also, envelopes that use linear segments with angled joints should be used with caution, if at all, as they often produce undesirable spectral components. Sometimes these 'undesirable' effects may be useful in order to add a certain coloration to the resulting sound, but as a rule of thumb they should be avoided.

Figure 5.9 Discontinuities between grains produce undesirable glitches. This may be prevented by using a suitable envelope

Despite the simplicity of a granular instrument, its operation requires the control of large sets of parameters. Each grain needs to be specified, but it is impracticable to specify these parameters manually. To begin with, it is very time consuming and difficult to abstract the overall outcome. It would be equivalent to having to manually write the values for thousands of samples in order to synthesise a sinewave. The sound designer is therefore urged to provide global grain parameters and to devise high-level mechanisms to control them. The global grain parameters commonly found in most granular synthesis programs are:

- Grain envelope shape
- Grain waveform
- Grain frequency
- Grain duration
- Delay time between grains
- Spatial location (e.g. in the stereo field)

High-level mechanisms to control these parameters are normally implemented using one or a combination of the following techniques:

- Trajectory envelopes
- Random values
- Probabilities
- Fractal geometry
- Cellular automata

In addition to global grain parameters and their control mechanisms, a granular synthesiser should also provide controllers at the level of the whole sound event, such as duration of the event. An example of controller that uses cellular automata is introduced below.

5.1.4 High-level granular control: a case study

On the accompanying CD-ROM (in the folder *chaosynth*), there is a demonstration version of Chaosynth, a sequential granular synthesis program for Macintosh and Windows. An introduction to Chaosynth is given in Chapter 8. In this section we will study the mechanism used to control its parameters.

The standard approach to granular synthesis control is to use probabilities to generate the parameter values for the production of the individual grains; e.g. one could set up the synthesiser to produce grains of 30 milliseconds duration with 50 per cent probability, 40-millisecond grains with 20 per cent and 50-millisecond grains with 30 per cent. Transition functions to change the probability as the sounds are produced can be defined in order to produce more dynamic sounds. Chaosynth uses a different method, which proved able to generate a much greater variety of interesting sounds than stochasticity: it uses *cellular automata*.

A brief introduction to cellular automata (CA) is given in the previous chapter where we introduced the cellular automata lookup table technique. Chaosynth, however, uses a two-dimensional arrangement of cells instead of a one-dimensional one. In general, two-dimensional CA are implemented as a regular matrix of cells. Each cell may assume values from a finite set of integers and each value is normally associated with a colour. The functioning of a cellular automaton is displayed on the computer screen as a sequence of changing patterns of tiny coloured cells, according to the tick of an imaginary clock, like an animated film. At each tick of the clock, the values of all cells change simultaneously, according to a set of transition rules that takes into account the values of their four or eight neighbours.

At a given moment t, the cells of Chaosynth's CA can be in any one of the following conditions: (a) *quiescent*, (b) in one of *n states of depolarisation* or (c) *collapsed*. The metaphor behind this CA is as follows: a cell interacts with its neighbours through the flow of electric current between them. There are minimum ($Vmin$) and maximum ($Vmax$) threshold values which characterise the condition of a cell. If its internal voltage (Vi) is under $Vmin$, then the cell is quiescent (or polarised). If it is between $Vmin$ (inclusive) and $Vmax$ values, then the cell is being depolarised. Each cell has a potential divider which is aimed at maintaining Vi below $Vmin$. But when it fails (that is, if Vi reaches $Vmin$) the cell becomes depolarised. There is also an electric capacitor which regulates the rate of depolarisation. The tendency, however, is to become increasingly depolarised with time. When Vi reaches $Vmax$, the cell fires and collapses. A collapsed cell at time t is automatically replaced by a new quiescent cell at time $t + 1$. The functioning of the CA is determined by a number of parameters, including: the number n of possible cell values or colours ($n >= 3$), the resistors $R1$ and $R2$ for the potential divider; the capacitance k of the electric capacitor and the dimension of the grid. These are Chaosynth user-specified settings.

In practice, the condition of a cell is represented by a number between 0 and $n - 1$ (n = amount of different conditions). A value equal to 0 corresponds to a quiescent condition, whilst a value equal to $n - 1$ corresponds to a collapsed condition. All values in between exhibit a degree of depolarisation. The closer the value gets to $n - 1$ the more depolarised the

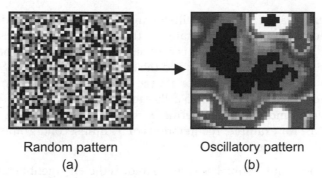

Random pattern   Oscillatory pattern
(a)   (b)

Figure 5.10 The CA of our system tend to evolve from an initial random distribution of cells in the grid (a) towards an oscillatory cycle of patterns (b)

cell becomes. The functioning of the automaton is defined by three rules simultaneously applied to each cell, selected according to its current condition. The rules are as follows:

- if **quiescent**: the cell may or may not become depolarised at the next tick of the clock ($t + 1$). This depends upon the number of polarised cells in its neighbourhood, the number of collapsed cells in its neighbourhood and the resistance of the cell to being collapsed
- if **depolarised**: the tendency is to become more depolarised as the clock t evolves
- if **collapsed**: a collapsed cell at time t generates a new quiescent cell at time $t + 1$

The automaton tends to evolve from an initial random distribution of cells in the matrix towards an oscillatory cycle of patterns (Figure 5.10).

Rendering sounds from cellular automata

Each sound grain produced by Chaosynth is composed of several spectral components. Each component is a waveform produced by a digital oscillator which needs two parameters to function: frequency (Hz) and amplitude (dB). In Chaosynth, the oscillators can produce various types of waveforms such as sinusoid, square, sawtooth, and band-limited noise. The CA control the frequency and duration values of each grain, but the amplitude values are set up by the user beforehand via Chaosynth's Oscillators panel. The system can, however, interpolate between two different amplitude settings in order to render *sound morphing* effects.

The mechanism works as follows: at each cycle, the CA produces one sound grain. In order to visualise the behaviour of a CA on the computer, each possible cell condition is normally associated with a colour, but in this case the system also associates these conditions to various frequency values. For example: yellow = 110 Hz, red = 220 Hz, blue = 440 Hz, and so forth; these are entirely arbitrary user-specified associations. Then, the matrix of the automaton is subdivided into smaller uniform sub-grids of cells and a digital oscillator is associated to each sub-grid (Figure 5.11). At each cycle of the CA, that is, at each snapshot of the CA 'animation', the digital oscillators associated with the sub-grids simultaneously produce signals, which are added in order to compose the spectrum of the respective granule. The frequency values for each oscillator are determined by the arithmetic mean over the frequency values associated to the states of the cells of their corresponding sub-grid. Suppose, for example, that each oscillator is associated with 9 cells and that at a certain cycle t, 3 cells correspond to 110 Hz, 2 to 220 Hz and the other 4 correspond to 880 Hz. In this case, the mean frequency value for this oscillator at time t will be 476.66 Hz. (An example of a grid of 400 cells allocated to 16 oscillators of 25 cells each is given in Figure 7.17, Chapter 7.) The duration of a whole sound event is determined by the number of CA iterations and the duration of the grains; for example, 100 iterations of 35 millisecond grains results in a sound event of 3.5 seconds of duration.

This mapping method is interesting because it explores the emergent behaviour of the CA in order to produce sounds in a way which resembles the functioning of some acoustic instruments. The random initialisation of cells in the grid produces an initially wide distribution of frequency values, which tend to settle to an oscillatory cycle. This behaviour resembles the way in which the sounds produced by most acoustic instruments evolve during their production: their harmonics converge from a wide distribution (as in the noise attack time of the sound of a bowed string instrument, for example) to oscillatory patterns (the characteristic of a sustained tone). Variations in tone colour are achieved by varying the

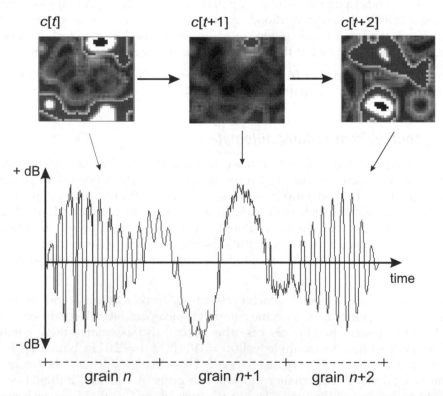

Figure 5.11 Each snapshot of the cellular automaton produces a sound grain. Note that this is only a schematic representation, as the grains displayed here do not actually correspond to these particular snapshots

frequency values, the amplitudes of the oscillators and the number of cells per oscillator. Different rates of transition, from noise to oscillatory patterns, are obtained by changing the values of $R1$, $R2$ and k.

5.2 Pulsar synthesis

Pulsar synthesis, as recently proposed by Curtis Roads (2001), shares the same fundamental principles of granular synthesis. The main difference is that the concept of sound grains is replaced by the notion of *pulsars*. In essence, a pulsar is a sound grain with a variable duty cycle; an idea inspired by an analog synthesis effect known as pulse-width-modulation (PWM). More specifically, a pulsar consists of a waveform w (or *pulsaret*) of duration d followed by a silent portion s (Figure 5.12). The period p of a pulsar is therefore given by the sum of $w + s$, and d corresponds to the duration of its duty cycle. The frequency of a stream of pulsars with period p is calculated as $1/p$ Hz.

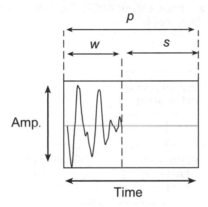

Figure 5.12 A pulsar consists of a waveform followed by a silent portion

In practice, a simple pulsar is implemented as a waveform that is limited in time by an envelope referred to as the pulsaret envelope. This envelope can be a variety of shapes and it has a strong influence on the timbre of the resulting sound. Typical pulsaret envelopes include rectangular, linear decay, linear attack and standard ADSR (Figure 5.13).

In principle, the nature of the waveform w can be anything from a sinewave to a band-limited pulse. Pulsars can also be used to convolve sampled sounds (Figure 5.14). By convolving pulsars with a brief sound sample, one imprints this sample onto the time pattern of the pulsars. The consequence is that the pulsars are then replaced by filtered copies of the sound sample.

A simple pulsar synthesiser will have at least the following control parameters:

- Duration of the pulsar stream: D
- The overall stream amplitude envelope: A
- The frequency of the pulsar stream: $F = 1/p$
- The waveform of the pulsaret: w

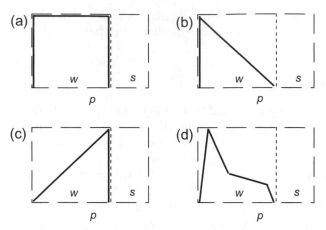

Figure 5.13 The duty cycle of a pulsar can be shaped by a number of different envelopes: (a) rectangular envelope, (b) linear decay, (c) linear attack and (d) ADSR. The nature of these envelopes has a strong influence on the timbre of the resulting sound

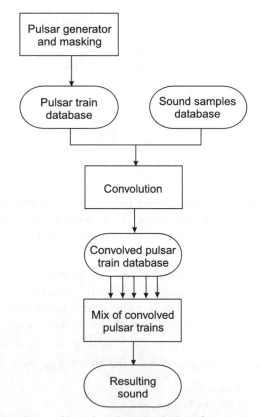

Figure 5.14 Pulsar synthesis scheme. Note that the convolution of a pulsar stream with a brief sound sample replaces the original pulsar on the stream by a filtered version of the sample

- The duration of the pulsaret: d
- The pulsaret envelope: v

The range of timbres that can be produced by pulsar synthesis is very large and its outcome is not straightfoward to predict. As with granular synthesis, perhaps the best way to work with pulsar synthesis is to experiment with various settings systematically until the desired result is obtained. The pulsaret envelope is perhaps the most significant controller in pulsar synthesis as it can change the resulting spectrum dramatically. The duration of the pulsaret is also very important, as it defines a sort of formant peak of frequency $1/d$ in the spectrum. Sweeping the parameter d often sounds similar to the well-known effect caused by a time-varying band-pass filter applied to a pulse train. Spectra with various resonant peaks can be produced by using as many pulsar generators as the number of peaks in synchrony with a common fundamental frequency F.

In the folder *pulsar* in the Macintosh portion of the accompanying CD-ROM there is a demo version of PulsarGenerator, a terrific pulsar synthesiser implemented by Alberto de Campo and Curtis Roads at the University of California, Santa Barbara.

5.3 Resynthesis by fragmentation and growth

Resynthesis by fragmentation and growth is akin to the granular sampling approach to granular synthesis, but its postulate is slightly different. Instead of producing huge quantities of short sounds to form larger sound events, this technique basically enlarges a short sound by repeating segments of itself (Figure 5.15).

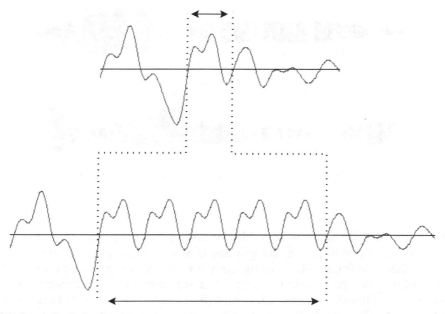

Figure 5.15 Resynthesis by fragmentation and growth basically works by enlarging a short sound by repeating segments of itself

Resynthesis by fragmentation and growth originates from one of the earliest analog electronic music composition techniques, known as 'tape looping'. Tape looping is a process whereby a short length of recorded tape is cut and both ends are spliced together to form a loop. A repetitive sound pattern is generated by playing back this loop. The music potential of this technique was extensively explored by composers of early electronic music.

Similar techniques are now available on some MIDI-samplers to simulate sustain on a note that is longer than the sampled sound itself. A number of computer programs for fragmentation and growth have been developed by the Composer's Desktop Project (CDP) in England (CDP is on the accompanying CD-ROM).

5.3.1 Brassage

Brassage is a technique whereby a sound is fragmented and spliced back in a variety of ways. If the fragments are spliced in the same manner as they were cut, then the original sound will be reproduced. But if, for example, the fragments are shuffled and overlaid, then a completely different, and often bizarre, result may be obtained (Figure 5.16).

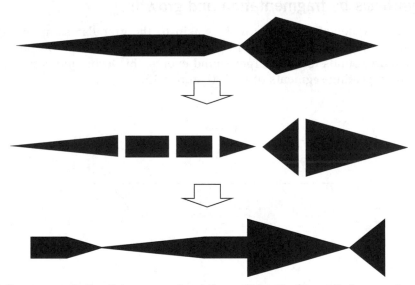

Figure 5.16 Brassage works by slicing a sound and then splicing the fragments in a number of different ways

The duration of a sound can be extended by cutting overlapping segments from the source but not overlapping them in the resulting sound (Figure 5.17). Conversely, the sound can be shortened by choosing fragments from the source with a space between them and splicing them back without the spaces (Figure 5.18). More sophisticated segmentation may involve the specification of different segment sizes and the variation of either or both the pitch and the amplitude of the fragments. These variations may, of course, be random or driven by a function or envelope.

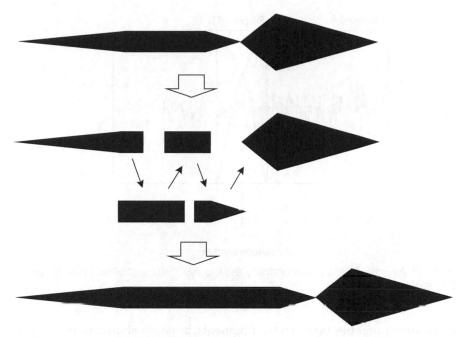

Figure 5.17 Brassage can extend the duration of a sound by slicing overlapping segments but not overlapping the segments in the resulting sound

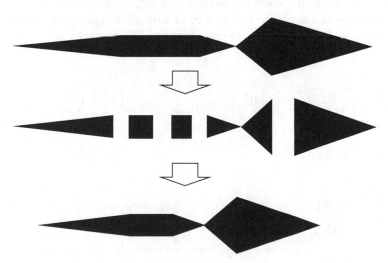

Figure 5.18 Brassage can shorten the duration of a sound by slicing segments with a space between them and then by splicing back the segments without the spaces

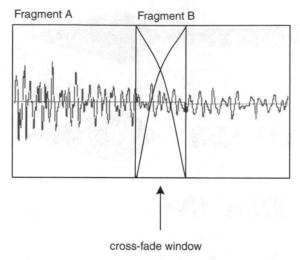

Figure 5.19 An overlap crossfade window is normally used to avoid discontinuities between the segments in the brassage process

In order to avoid discontinuities between the fragments, brassage should normally join them together using an overlapping crossfade window (Figure 5.19). If the fragment sizes are very short and uniform, then the crossfading may cause a similar envelope modulation effect as with granular synthesis. In most cases an exponential crossfade curve produces better splicing.

5.3.2 Zigzag

Zigzag is a process in which a sound is played in a zigzag manner. The reversal points can be specified anywhere in the sound and can be of variable length. If the process starts at the beginning of the sound and stops at its end (Figure 5.20), then the result will appear to be

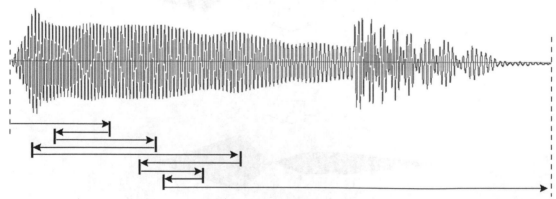

Figure 5.20 The reversal points of a zigzag process can be specified anywhere in the sound and can be of variable length

artificially extended. The aural effect will depend upon the nature of the original sound. The extension effect will be more convincing if zigzag is applied to highly homogeneous or very 'noisy' sounds.

Interesting results may be obtained with controlled changes in length and position, thus producing 'undulatory' sound effects. As in brassage, zigzag also requires the careful specification of a crossfade window to smooth the continuation of the playback. Again, envelope modulation is bound to occur if the zigzag span is too short and too fast.

An abridged version of CDP's ZIGZAG program is available on the CD-ROM (in the folder *cdpdemo*). Use a Web browser to read the user manual (*Cdpman.htm*) for more information about ZIGZAG.

5.4 Waveset distortion

Waveset distortion involves the transformation of a sound at its wavecycle level; that is, the unit for waveset distortion is a set of samples between two successive zero crossings. For example, a simple sinusoid sound is viewed in this context as a succession of wavecycles, and not as a succession of samples.

As for resynthesis by fragmentation and growth, waveset distortion is akin to the granular sampling approach to granular synthesis. The main difference here is that radical changes are applied to the individual wavecycles, rather than mere substitution of either the order or location of grains. Also, there is an explicit unit definition to work with here – the wavecycle – as opposed to granular synthesis, where the size of the grain is an implicit subjective unit. Note that in this case the concept of pitch does not always apply. For a simple periodic sound, the greater the number of wavecycles, the higher the pitch, but for a more 'wobbly' sound this may not be the case.

The composer Trevor Wishart has developed a number of waveset distortion programs for the CDP package. There are two types of program in Wishart's kit: programs that work on individual wavecycles and those that work on groups of wavecycles (that is, wavesets). Typical waveset distortion operations are:

- Addition or removal of wavecycles in a waveset
- Modulation of wavecycles or wavesets
- Application of amplitude envelopes to wavecycles
- Replacement of wavesets by other waveforms (e.g. sine, square)
- Permutation of wavecycles

There are three CDP programs for waveset distortion available on the CD-ROM: RESHAPE, OMIT and INTERACT. Whilst RESHAPE reads each wavecycle and replaces it with a different waveform of the same length, the OMIT program adulterates a sound file by omitting wavecycles, thus producing holes of silence in the sound. As for INTERACT, it alternately interleaves the wavecycles of two different sounds. Use a Web browser to read the CDP user manual (file *Cdpman.htm*) for more information on these programs.

5.5 PSOLA

PSOLA, an acronym for Pitch Synchronous Overlap and Add, was originally designed for speech synthesis. In short, it works by concatenating small segments with an overlapping factor. The duration of these segments is proportional to the pitch period of the resulting signal. PSOLA is useful for composers because a given signal can be decomposed in terms of these elementary segments and then resynthesised by streaming these elements sequentially (Figure 5.21). Parametrical transformation can be applied during the streaming of the segments in order to modify the pitch and/or the duration of the sound. PSOLA is particularly efficient for shifting the pitch of a sound, as it does not, in principle, change its spectral contour. PSOLA closely resembles brassage and waveset distortion dicussed earlier, with the difference that the segmentation is much more constrained here.

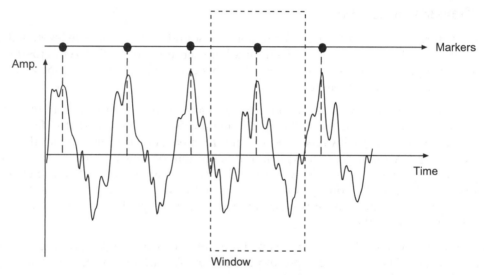

Figure 5.21 The PSOLA analysis and resynthesis technique first decomposes a signal into small overlapping segments, called markers, using a window whose middle point is the segment's point of maximum energy. The sound is then resynthesised by streaming these segments. Duration and pitch changes can be made during resynthesis by changing the original disposition of the markers

At the analysis stage, the elementary segments are extracted using a window centred at the local maxima, called *markers*. These markers should be pitch-synchronous in the sense that they should lie close to the fundamental frequency of the sound. The size of the window is proportional to the local pitch period. The pitch of the signal can be changed during re-synthesis by changing the distances between the successive markers (Figure 5.22). Time stretching or compression is achieved by repeating or skipping segments (Figure 5.23).

The caveat of the PSOLA analysis and resynthesis method is that it only works satisfactorily on sounds containing regular cycles displaying a single energy point, referred to as the *local maxima*; the human voice fulfils this requirement, but not all sounds do. The good news is that PSOLA is fast, as it deals directly with the sound samples in the time-domain.

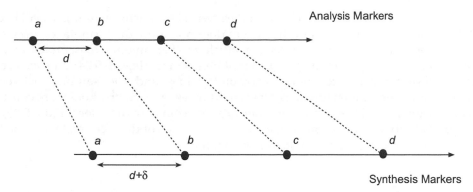

Figure 5.22 Pitch-shift in PSOLA by changing the distances between the successive markers. In this example the pitch is lowered by enlarging the distance *d* by a factor *d* + δ

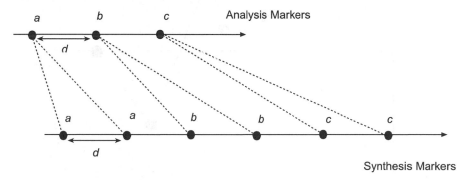

Figure 5.23 Time-stretching in PSOLA by repeating segments

The Praat system on the accompanying CD-ROM (in the folder *praat*) features an efficient PSOLA tool. Please refer to the user manual which is accessed via the program's Menu panel for instructions on how to use the PSOLA tool.

5.6 Statistical wavecycle synthesis

This technique works by synthesising waveforms from a sequence of interpolated breakpoint samples produced by a statistical formula; e.g. Gaussian, Cauchy or Poisson. The interpolation may be calculated using a number of functions, such as exponential, logarithmic and polynomial. Due to the statistical nature of the wavecycle production, this technique is also referred to as *dynamic stochastic synthesis*.

This technique was developed by the composer Iannis Xenakis at Cemamu (Centre for Research in Mathematics and Music Automation) in France. Xenakis also designed a synthesis system called Gendy (Serra, 1993). Gendy's primary assumption is that no wavecycle of a sound should be equal to any other, but rather that it should be in a constant process of transformation. The synthesis process starts with one cycle of a sound and then modifies its form each time the cycle is synthesised. The transformation is calculated by a probability distribution formula but the system does not apply the formula to every sample of the wavecycle. Instead, Gendy selects a much smaller amount of samples to forge a representation of the wavecycle as a sequence of breakpoints (Figure 5.24). At each cycle, the probability formula calculates new breakpoint values and the samples between are generated by an interpolation function. New coordinates for each breakpoint are constrained to remain within a certain boundary in order to avoid complete disorder (Figure 5.25). Also, different probability functions can be specified for each axis of the coordinate; i.e. one for the amplitude and one for the position of the breakpoint.

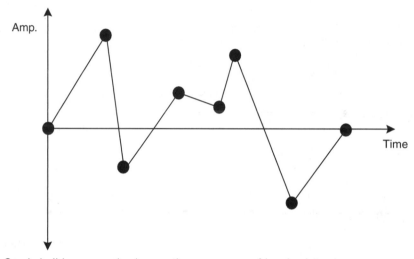

Figure 5.24 Gendy builds wavecycles by creating sequences of breakpoint values

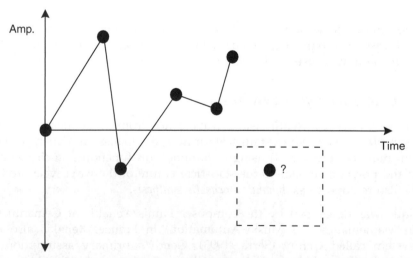

Figure 5.25 In order to avoid complete disorder, new breakpoint values are generated using a probability function that narrows the result to a certain range of values

The parameters for statistical wavecycle synthesis are normally:

- Number of breakpoints in a wavecycle
- Type of probability function(s)
- Type of interpolation
- Boundary constraints for each breakpoint of the wavecycle

5.7 Sequential waveform composition

Sequential waveform composition works by letting the computer generate sample pattern sequences according to a user-defined formula. The patterns need not be a single wavecycle or a waveset. In fact, the whole sequence would not necessarily need to form any pattern at all. Some patterning is, however, desirable here because a minimum degree of repetition is indispensable for the production of a waveform. In other words, repetition is necessary because it determines the 'form' of a waveform, otherwise the outcome will always be white noise.

In essence, the technique resembles the statistical wavecycle synthesis technique introduced earlier, with the fundamental difference that here the 'breakpoints' are deterministic, in the sense that the segments are defined explicitly rather than statistically. Despite the deterministic nature of the technique itself, its outcome is often unpredictable, but notwithstanding interesting and unique. Like the binary instruction technique discussed in Chapter 2, synthesis by sequential waveform composition is also commonly referred to as *non-standard synthesis*. Indeed, in order to work with this technique one needs to be prepared to work with non-standard ways of thinking about sounds, because its synthesis parameters by no means bear a direct relation to acoustics.

The composer Arun Chandra, based in Illinois, USA, is one of the main supporters of the sequential waveform composition technique (Chandra, 1994). He has devised Wigout and TrikTrak, a unique pair of programs for PC-compatible platforms, specifically designed for sequential waveform composition. Both programs, plus a number of examples and a comprehensive tutorial – specially prepared for this book by Arun Chandra himself – are available on the accompanying CD-ROM (in the folder *wigtrik*).

In order to understand the fundamental principles of this technique one should look at the basics of digital sound synthesis introduced in Chapter 1 from a slightly different angle. To reiterate, in order to produce a sound the computer feeds a series of numbers, or samples, to a digital-to-analog converter (DAC). The range of numbers that a DAC can manage depends upon the size of its word (refer to Chapter 1). For example, a 16-bit DAC handles 65 536 different values, ranging from −32 768 to +32 767 (including zero). The number of sample values the DAC converts in one second is referred to as the sampling rate. For instance, a sampling rate equal to 44 100 Hz means that the DAC converts 44 100 samples (i.e. numbers between −32 768 and +32 767) per second.

In order to produce musically useful sounds on a computer, it must be programmed to produce the right sequence of samples. If the computer keeps producing the same sample value (e.g. 20 000), the result will be silence (Figure 5.26). Conversely, if the computer is programmed to produce sequences of different streams of sample values, then the output

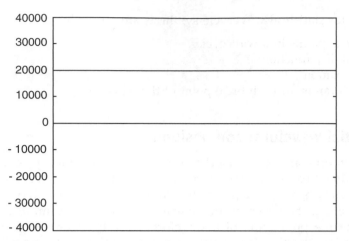

Figure 5.26 If the computer produces the same sample all the time, the result will be silence

will be audible. It is exactly at this level that Wigout and TrikTraks work. For example, suppose that the computer is set to produce sequences of four different sample streams, as follows:

- A stream of 20 samples equal to +30 000 followed by
- A stream of 30 samples equal to −20 000 followed by
- A stream of 10 samples equal to +16 000 followed by
- A stream of 40 samples equal to −5000

In this case, the computer produces the waveform depicted in Figure 5.27. The whole sequence (referred to as a *state*, in Wigout terms) is composed of 100 samples in total. Thus in order to produce a one-second sound at a sampling rate of 44 100 Hz, the computer must repeat this 'state' 441 times (441 3 100 = 44 100).

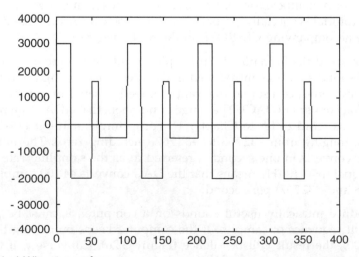

Figure 5.27 A typical Wigout waveform

In Wigout and TrikTrak terms, each stream of homogeneous samples is referred to as an *element*. In the case of Figure 5.27, the waveform is specified in terms of four elements; each element is specified by two parameters: amount of samples (or 'duration') and amplitude (in this case a value between −32 768 and +32 767):

- Element 1: a stream of 20 samples equal to +30 000
- Element 2: a stream of 30 samples equal to −20 000
- Element 3: a stream of 10 samples equal to +16 000
- Element 4: a stream of 40 samples equal to −5000

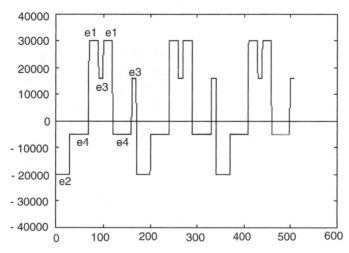

Figure 5.28 A variation of the waveform depicted in Figure 5.27

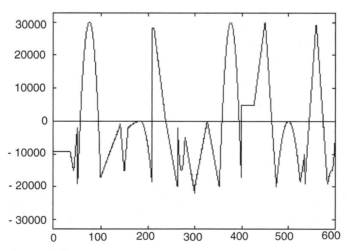

Figure 5.29 Besides squared shapes, Wigout also supports triangular shapes, curved shapes and a combination of all three

Essentially, the technique works by specifying a collection of elements and by arranging these elements in a sequence that together comprises the waveform; hence the name: *sequential waveform composition*. A variation of the waveform portrayed in Figure 5.27 could be defined using the same four elements as follows (Figure 5.28): element 2 + element 4 + element 1 + element 3 + element 1 + element 4 + element 3.

A great variety of waveforms can be produced using this technique. Wigout allows for the specification of waveforms that change the duration and the amplitude of their elements during production. Also, besides squared shapes one can also define triangular shapes, curved shapes and a combination of all three (Figure 5.29). TrikTraks complements the functionality of Wigout by adding the means for the specification of more complex waveform transitions; e.g. using non-linear functions.

6 Practical case studies and sound design secrets: from humanoid singing to Klingon phasers

One of the great strengths of computer sound synthesis is its potential for the creation of an infinite variety of instruments and sounds. So far this book has introduced a number of synthesis techniques to explore this potential. Ideally, musicians should master all these techniques, but this is neither practical nor absolutely necessary for the production of good music. In practice, composers have an idea of what is generally possible and then select a few techniques on which to build their own modus operandi.

When designing synthesis instruments, we should give preference to those techniques that best fit the way we understand the sound phenomena we want to work with. For example, whilst those composers interested in timbral manipulation will be better off working with analysis and resynthesis techniques (Chapter 3), those working with scientific metaphors (e.g. dynamic systems, cellular automata, etc.) would certainly prefer to use granular or statistical synthesis methods (Chapter 5).

It is impossible to develop a set of definite rules for sound and instrument design, but we can, however, establish some guidelines. These should normally emerge from the study of the work of experienced composers, and, of course, from your own experience.

This chapter introduces a few case studies of instrument and sound design. It begins with a discussion on the relationship between the sound phenomena and synthesis techniques. Then, an approach to instrument and sound design is outlined. This should by no means be considered as a finite set of rules to be followed strictly, but rather as a reference for the outline of your own approach. Next we study the task of synthesising the human voice in greater detail. A step-by-step tutorial on building a subtractive formant synthesiser is followed by an introduction of how to control Praat's physical modelling voice system (available on the accompanying CD-ROM; refer to Chapter 8). Finally we introduce a discussion on devising new taxonomies for categorising granular synthesis sounds.

6.1 Associating cause and effect

Unless you are prepared to adopt a highly irrational approach to sound design, you should always give preference to basing your new instruments on well-documented acoustic theories; see Howard and Angus (1996) for a good introduction to acoustics. Acoustic musical instruments are a reliable point of departure because the human ear is familiar with them. However, be wary here, because the expression 'point of departure' does not necessarily mean that you should always attempt to imitate these instruments. Bear in mind that software sound synthesis gives you the power to create new instruments and sonorities.

The basic spectral formation of the sounds produced by acoustic musical instruments may be classified into three general categories: sounds with *definite pitch*, sounds with *indefinite pitch* and *noise*. The analysis of the spectrum of these sounds reveals that:

- Sounds with definite pitches often have harmonic partials
- Sounds with indefinite pitches often have inharmonic partials
- Noises tend to contain a large number of random components covering the whole audible band

This is indeed a very simplistic way of categorising sounds. Nevertheless, it can be a good starting point if you can fit the sounds you want to synthesise into one of these three categories. This will help you to establish which techniques will be more appropriate to produce them. For example:

- Sawtooth and narrow pulses are good starting points to model the harmonic partials produced by vibrating strings and open air columns; e.g. strings, brass
- Square and triangular waves are good starting points to model sounds with only odd harmonic partials produced by closed air columns; e.g. woodwinds, organ
- Amplitude and frequency modulations are good starting points to model the non-hamonic partials of sounds produced by hanging bars, bells and drums
- Noise generators are the best starting point to model the highly random partial configurations of the sounds of winds, snare drums and air blasts through a cylinder

Also, the changes of the 'shape' of a sound event over its duration are vital clues for the identification of the acoustic mechanisms that could have produced the sound. Acoustic instruments are played by an excitation which sets it into vibration. The type of excitation defines the shape of the sound; that is, whether the sound is plucked, bowed or blown, for example. There are two general ways of producing vibrations on an acoustic instrument: by *continuous excitation* or by *impulsive excitation*.

Continuous excitation is when the instrument needs the constant application of force to produce a note. In this case the sound will be heard as long as the force is being applied (e.g. a clarinet). Sounds from continuous excitation-type instruments often have a relatively slow attack time and will sustain the loudness until the excitation stops. Most sounds produced by continuous excitation begin with a very simple waveform at its attack time. Then, an increasing number of partials appear and build up the spectrum until the sound reaches the sustain time. When the excitation ceases, the partials often fade out in the opposite order to that in which they appeared.

Impulsive excitation is when the instrument requires only an impulse to produce a note. In this case, the attack time is very short, almost instantaneous. The sound begins at its loudest and fullest instant, but it has virtually no sustain. As the sound fades, its spectrum normally fades away in a descending order.

In his *Traité des objets musicaux* Pierre Schaeffer (1966) purports to define criteria for the identification and classification of sounds according to their general features. Although Schaeffer was not explicitly concerned with sound synthesis itself, his work provides an excellent insight into making sound categorisations. One of the most important aspects of his theoretical work is the concept of *maintenance*, which involves the relationship between both criteria discussed above: the *basic spectral formation* and *mode of excitation*.

Schaeffer observed that sounds result from a certain energetic process, called *maintenance*, which describes the development that a sound undergoes with time. If the sound is merely ephemeral, a non-resonant single sound such as a drum stroke or a vocal plosive consonant,

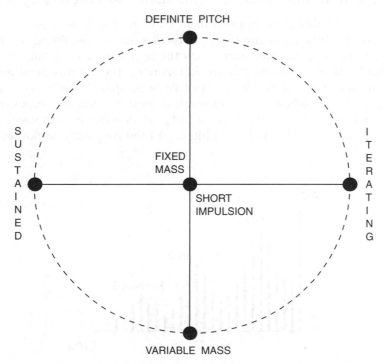

Figure 6.1 The two-axis diagram inspired by Pierre Schaeffer's concept of sound maintenance is an excellent example of a framework for sound categorisation

then there is a discrete short impulsion. If the sound is prolonged, such as a sung vowel, then there is a continuous, that is, sustained, sound. If it is prolonged by the repetition of impulsion, such as a drum roll, then there is iterating maintenance.

For a more systematic panorama, a two-axis diagram (Figure 6.1) can be defined. The middle horizontal axis contains sounds of *short impulsion*, the left features the *sustained sounds* and the right has those whose maintenance is *iterating*. On the vertical axis, sounds with *fixed mass* are placed between sounds of *definite pitches* and sounds of *variable mass*. Most sounds produced on acoustic musical instruments settle extremely fast into a fixed pitch or fixed mass (e.g. pink noise), following the attack stage. *Mass*, in this case, can be considered as the amount of information carried by the sound, ranging from redundancy (a meagre steady sinewave of fixed pitch) to total unpredictability (white noise). These perceptual phenomena may be implemented in a number of ways; for example by inputting noise into a band-pass filter whose passband is shortened (this tends to produce a sound with definite pitch) and widened (which results in a sound with a variable noise band).

The two-axis diagram in Figure 6.1 is an example of a starting point for the design of synthesis instruments. It provides an excellent framework for sound categorisation and encourages the specification of meaningful synthesis parameters; e.g. the bandwidth of the BPF mentioned above controls the mass of the sound.

6.2 Synthesising human vocal sounds using subtractive synthesis

This section studies the design of an instrument that could well be used to implement the vocal mechanism of a humanoid singer: it produces sounds resembling the human voice. The spectral contour of vocal-like sounds has the appearance of a pattern of 'hills and valleys' technically called formants (Figure 6.2). Among the various synthesis techniques capable of synthesising formants, the subtractive technique (Chapter 4) is used for this example. In subtractive synthesis each formant is associated with the response of a BPF; in this case, one needs a parallel composition of BPFs set to different responses. The signal to be filtered is simultaneously applied to all filters and the frequency responses of the filters are added together.

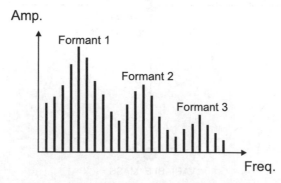

Figure 6.2 In subtractive synthesis each formant is associated with the response of a band-pass filter (BPF)

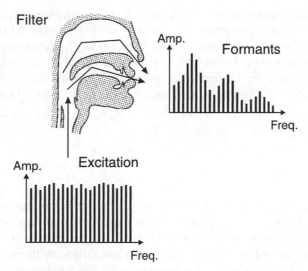

Figure 6.3 The subtractive synthesis of formants is determined by two main components: excitation source and resonator. The former produces a raw signal that is shaped by the latter

6.2.1 The glottal source component

In Chapter 4 we introduced the physiology of the human vocal mechanism from a physical modelling point of view. Here we will study the vocal tract from a subtractive synthesis standpoint.

When singing or speaking, an airstream is forced upwards through the trachea from the lungs. At its upper end, the trachea enters the larynx, which in turn opens into the pharynx. At the base of the larynx, the vocal folds are folded inwards from each side, leaving a variable tension and a slit-like separation, both controlled by muscles in the larynx. In normal breathing, the folds are held apart to permit the free flow of air. In singing or speaking, the folds are brought close together and tensed. The forcing of the airstream through the vocal folds in this state sets them into vibration. As a result, the airflow is modulated at the vibration frequency of the vocal folds. Despite the fact that the motion of the vocal folds is not a simple but a non-uniform vibration, the pitch of the sound produced is determined by this motion. In order to simulate this phenomenon, our subtractive instrument generates two types of excitation sources: the *voicing excitation source*, which produces a quasi-periodic vibration, and the *noise excitation source*, which produces turbulence. The former generates a pulse stream intended to simulate the non-uniform (or quasi-periodic) vibration of the vocal folds, whereas the latter is intended to simulate an airflow past a constriction or a relatively wide separation of the vocal folds.

The voicing excitation source

Figure 6.4 portrays the voicing excitation source mechanism of our instrument. *Jitter* and *vibrato* are very important for voicing sound quality because they add a degree of non-uniformity to the fundamental frequency of the excitation. Jitter is defined as the difference in fundamental frequency from one period of the sound to the next. It normally varies at random between –6 per cent and +6 per cent of the fundamental frequency. Our instrument calculates this value by adding the results from three random number generators whose values are produced by interpolating periodic random values. The pcmusic code for the jitter generator is as follows (refer to Chapter 8 for more informtation about pcmusic):

```
ran b10 0.02 0.05 Hz d d d;
ran b11 0.02 0.111 Hz d d d;
ran b12 0.02 1.219 Hz d d d;
adn b13 b10 b11 b12;
mult b14 b13 F0;              // b14=jitter factor
```

Vibrato is defined as the frequency modulation of the fundamental frequency. In general, vibrato is interesting from a timbral point of view. For instance, in humans it is important for the recognition of the identity of the singer. Typically, there are two distinct parameters for vibrato control: the *width* of the vibrato, that is, the amplitude of the modulating frequency, and the *rate* of the vibrato (VIBR), the frequency of the modulating frequency. VIBR is normally set to a value between 5.2 Hz and 5.6 Hz. If this value is set above the lower limit of the audio range (approximately 18 Hz), then the effect will no longer be a vibrato but a distortion (i.e. FM) – which may be useful to produce some interesting effects. Vibrato is implemented using an oscillator and is used in conjuction with jitter to produce variations in the fundamental frequency, represented here as F0:

```
osc b15 0.26 VIBR f2 d;      // b15=vibrato
adn b16 b15 b14 F0;          // b16=freq+jitter+vibrato
```

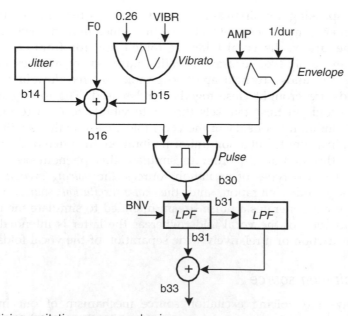

Figure 6.4 The voicing excitation source mechanism

The heart of the voicing excitation source is a pulse generator. It produces a periodic waveform at a specific frequency with a great deal of energy in the harmonics. A pulse waveform has significant amplitude only during relatively brief time intervals, called pulse width (Figure 6.5). When a pulse waveform is periodically repeated, the resulting signal has a rich spectrum. The output of the voicing excitation source provides the raw material from which the filtering system will shape the required sound.

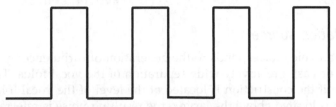

Figure 6.5 The spectrum of a pulse waveform has both odd and even harmonics

A human singer manages the source excitation signal by adjusting the muscles of the larynx, controlling the spacing and tension of the vocal folds. In loud singing the vocal folds close completely for a significant fraction of the vibration cycle. The resulting airflow is normally referred to as the *normal voicing* and it has a waveform like the one portrayed in Figure 6.6. If the folds are too slack, they do not close properly at any point in the vibration cycle. Here the resulting waveform loses its spiky appearance and starts to resemble a sinusoidal curve; this signal is referred to as the *quasi-sinusoidal* voicing. This results in the breathy sound quality often found among untrained singers. The expression 'trained singer' here denotes one with the ability to maintain the vocal folds pressed tightly together.

In order to simulate the production of both types of excitation signals, the pulse train is sent through two LPFs in series. The output from the first filter in the chain falls off at approximately 12 dB per octave above the cut-off low-pass frequency, thus simulating the normal voicing signal depicted in Figure 6.6. The cut-off frequency for this filter is normally set to an octave above the value of the fundamental frequency. Next, the output of the second LPF should produce a smoother quasi-sinusoidal voicing if its cut-off frequency is set to approximately one octave above the value of the cut-off of the first. Some degree of quasi-sinusoidal voicing can be added to the normal voicing source, in combination with aspiration noise (see below), in order to produce the above-mentioned breathy sound quality.

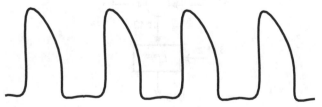

Figure 6.6 The simulation of a voicing source signal may be obtained by low-pass filtering the pulse train

The pcmusic code for this voicing mechanism is given as follows:

```
osc b20 AMP PER f1 d;              // b20=envelope voicing
blp b30 b20 b16 b16 30 d d;        // b30=pulse
NRES(b31, b30, 0 dB, 0, BNV);      // b31=LPFed normal
osc b20 AMP PER f1 d;              // voicing
NRES(b32, b31, -6 dB, 0, BQS);     // b32=LPFed quasi-
osc b20 AMP PER f1 d;              // sinusoidal
adn b33 b31 b32;                   // b33=voice source
```

The noise excitation source

Noise in the human voice corresponds to the generation of turbulence by the airflow past a constriction and/or past a relatively wide separation of the vocal folds. The resulting noise is called *aspiration* if the constriction is located at the level of the vocal folds; i.e. the larynx. If the constriction is located above the larynx, the resulting noise is referred to as *fricative*. If it is produced by a relatively wide separation of the vocal folds then the resulting noise is called *bypass*.

Our instrument produces noise using a random number generator, an offset modulator, and a LPF (Figure 6.7). It generates only aspiration noise; fricative and bypass are left out for the sake of the simplicity of this study. The pcmusic code for the noise source is given below:

```
osc b21 ASP PER f1 d;              // b21=envelope
osc b21 ASP PER f1 d;              // aspiration
white b40 b21;
osc b41 ASP b16 f3 d;              // noise modulator
adn b42 b40 b41;
NRES(b43, b42, 0 dB, 0, 22624 Hz);  // b43=noise source
```

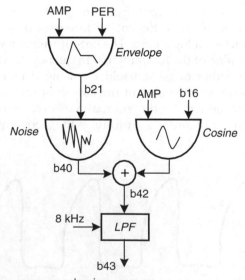

Figure 6.7 The noise excitation source mechanism

6.2.2 The resonator component

On its journey through the vocal tract the excitation signal is transformed. Components which are close to one of the resonance frequencies of the tract are transmitted with high amplitude, while those which lie far from a resonance frequency are suppressed. Much of the art of the singer lies in shaping the vocal tract in such a way that the crude source output is transformed into a desired sound. The vocal tract can be thought of as a pipe from the vocal folds to the lips plus a side-branch leading to the nose, with a cross-section area which changes considerably.

The length and shape of a particular human vocal tract determine the resonance in the spectrum of the voice signal. The length of the human vocal pipe is typically about 17 cm, which can be slightly varied by raising or lowering the larynx and shaping the lips. The cross-sectional area of the pipe is determined by the placement of the lips, jaw, tongue and velum. For the most part, however, the resonance in the vocal tract is tuned by changing its cross-sectional area at one or more points. A variety of sounds may be obtained by adjusting the shape of the vocal tract during phonation.

Five band-pass filters are appropriate for simulating a vocal tract with a length of about 17 cm (Figure 6.8). A typical female vocal tract is 15–20 per cent shorter than that of a typical male. In this case, four filters only are enough to simulate a female vocal tract. Each filter introduces a peak in the magnitude spectra determined by the passband centre frequency and by the formant bandwidth. The passband centre frequencies and bandwidths of the lowest three formants vary substantially with changes in articulation, whereas the fourth and fifth formants do not vary as much.

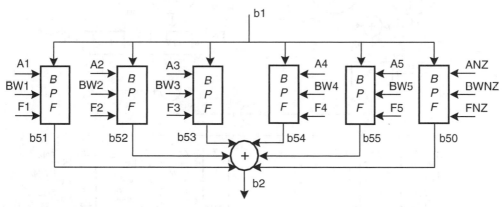

Figure 6.8 The resonator component of the example instrument is composed of six band-pass filters

When singing vowels, the connecting passage between the throat and the nasal cavity should be closed by lifting the velum. Opening this passage while singing produces a sound usually described as *singing through the nose*. The resonance of the nasal cavity is of great importance for the production of some consonants. The side-branch leading to the nasal cavity may be roughly simulated by introducing a 'nasal' sixth BPF. The pcmusic code for the resonator component is given as follows:

```
adn b1 b33 b43;                  // b1=voice+noise
NRES(b50, b1, ANZ, FNZ, BWNZ);   // b50=nasal formant
NRES(b51, b1, A1, F1, BW1);      // b51=formant 1
NRES(b52, b1, A2, F2, BW2);      // b52=formant 2, etc.
NRES(b53, b1, A3, F3, BW3);
NRES(b54, b1, A4, F4, BW4);
NRES(b55, b1, A5, F5, BW5);
adn b2 b51 b50 b52 b53 b54 b55;  // b2=whole formant
```

6.2.3 The full instrument and its pcmusic programming code

Figure 6.9 depicts the architecture of our instrument and its pcmusic score is given below; this instrument is available on the accompanying CD-ROM, (in the folder *Chapt6*) within pcmusic's materials. Note the extensive use of the *#define* instruction to create symbolic labels

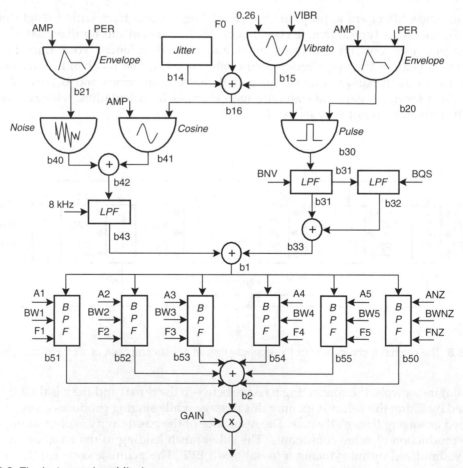

Figure 6.9 The instrument architecture

for the various synthesis parameters of the instrument, some of which represent variables whose values are provided in the note list (e.g. F0, AMP, etc.), whereas others represent constant values (F1, BW1, A1, etc.). The resonators, for example, are set with constant values to produce the vowel /a/, as in the word 'car' in English. (The values to produce the formants of other vowels can be found in Appendix 2.)

Also, the #*define* instruction is used to define the macro SING, which is a shorthand to facilitate the specification of note lists. Each SING statement needs six parameters: start time (in seconds), duration (in seconds), voicing source attenuation (in dB), aspiration noise attenuation (in dB), pitch (in Hz) and vibrato rate (in Hz), respectively.

```
// ------------------------------------------------
#include <cmusic.h>
#include <waves.h>
// ------------------------------------------------
#define F0 p7                   // fundamental frequency
#define AMP p5                  // amplitude voicing
#define ASP p6/2                // amplitude noise
#define VIBR p8                 // vibrato rate
#define PER 1/p4 Hz             // period for envelope
#define F0 p7                   // oscillator
#define BNV (p7*2)*1.414        // LPF voice source → normal
#define F0 p7                   // voicing
#define BQS (p7*4)*1.414        // LPF normal voicing →
#define F0 p7                   // quasi sinusoidal
#define F1 622.25 Hz            // formant 1 centre frequency
#define BW1 60 Hz               // formant 1 bandwidth
#define A1 0 dB                 // formant 1 attenuation
#define F2 1046.5 Hz            // formant 2 centre frequency
#define BW2 70 Hz               // formant 2 bandwidth
#define A2 -7 dB                // formant 2 attenuation
#define F3 2489 Hz              // formant 3 centre frequency
#define BW3 110 Hz              // formant 3 bandwidth
#define A3 -9 dB                // formant 3 attenuation
#define F4 2900 Hz              // formant 4 centre frequency
#define BW4 130 Hz              // formant 4 bandwidth
#define A4 -12 dB               // formant 4 attenuation
#define F5 3250 Hz              // formant 5 centre frequency
#define BW5 140 Hz              // formant 5 bandwidth
#define A5 -22 dB               // formant 5 attenuation
#define FNZ 311.12 Hz           // nasal formant centre
#define F0 p7                   // frequency
#define BWNZ 70 Hz              // nasal formant bandwidth
#define ANZ -22 dB              // nasal formant attenuation
#define GAIN 10                 // output gain
ins 0 voice;
```

```
// -----Voice source--------------------------------
// -----Jitter -------------------------------------
ran b10 0.02 0.05 Hz d d d;
ran b11 0.02 0.111 Hz d d d;
ran b12 0.02 1.219 Hz d d d;
adn b13 b10 b11 b12;
mult b14 b13 F0;                        // b14=jitter factor

// Vibrato ------------------------------------------
osc b15 0.26 VIBR f2 d;                 // b15=vibrato
adn b16 b15 b14 F0;                     // b16=freq+jitter+vibrato


// Voicing ------------------------------------------
osc b20 AMP PER f1 d;                   // b20=envelope voicing
blp b30 b20 b16 b16 30 d d;             // b30=pulse
NRES(b31, b30, 0 dB, 0, BNV);           // b31=LPFed normal
                                        // voicing
NRES(b32, b31, -6 dB, 0, BQS);          // b32=LPFed quasi-
osc b20 AMP PER f1 d;                   // sinusoidal
adn b33 b31 b32;                        // b33=voice source


// Noise source--------------------------------------
osc b21 ASP PER f1 d;                   // b21=envelope
osc b20 AMP PER f1 d;                   // aspiration
white b40 b21;
osc b41 ASP b16 f3 d;                   // noise modulator
adn b42 b40 b41;
NRES(b43, b42, 0 dB, 0, 22624 Hz);      // b43=noise source


// Resonators ---------------------------------------
adn b1 b33 b43; // b1=voice+noise
NRES(b50, b1, ANZ, FNZ, BWNZ);
                                        // b50=nasal formant
NRES(b51, b1, A1, F1, BW1);             // b51=formant 1
NRES(b52, b1, A2, F2, BW2);             // b52=formant 2, etc.
NRES(b53, b1, A3, F3, BW3);
NRES(b54, b1, A4, F4, BW4);
NRES(b55, b1, A5, F5, BW5);
adn b2 b51 b50 b52 b53 b54 b55;         // b2=whole formant


// Output -------------------------------------------
mult b2 b2 10;
out b2;
end;
```

```
GEN4(f1) 0, 0 -5 0.2, 1 1 0.8, 0.7 -5 1, 0;
SINE(f2);

COS(f3);

#define SING(st,dur,amp,aspir,freq,vibr)\
note st voice dur amp aspir freq vibr

SING(0.00, 0.55, -2 dB, -28 dB, 196.00 Hz, 5.2 Hz);
SING(0.55, 0.55, -4 dB, -30 dB, 155.56 Hz, 5.2 Hz);
SING(1.12, 3.79, -6 dB, -33 dB, 138.81 Hz, 5.4 Hz);
SING(5.05, 0.50, -3 dB, -31 dB, 174.61 Hz, 5.3 Hz);
SING(5.55, 2.00, 0 dB, -32 dB, 196.00 Hz, 5.3 Hz);
SING(8.02, 1.98, -3 dB, -30 dB, 155.56 Hz, 5.2 Hz);
SING(10.00, 0.60, -3 dB, -30 dB, 138.81 Hz, 5.3 Hz);
SING(10.62, 2.60, 0 dB, -32 dB, 174.61 Hz, 5.3 Hz);
ter;
```

An identical synthesiser implemented in Nyquist (Chapter 8) is available in the folder *voice* within the Nyquist materials on the accompanying CD-ROM; the file is called *articulator.lsp* and it is inside the folder *voicesynth*. Essentially, it is the same synthesiser furnished with a geometric articulator whereby the user specifies the rounding of the lips and the position of the tongue, instead of formant values. In short, this geometric articulator is implemented as a set of functions that estimates the formant values associated with the rounding of the lips, and the horizontal and vertical position of the body of the tongue in relation to the centre of the mouth.

6.3 Physical modelling synthesis of the human voice

This section presents an introductory tutorial on how to control the physical model of the human vocal system that comes with the Praat system (Chapter 8). Praat was originally designed as a tool for research in the field of phonetics, but it features a sophisticated physical modelling vocal synthesiser (referred to as *articulatory synthesis*) that can produce vocal and vocal-like sounds of great interest to composers and sound artists. A brief introduction to the architecture of this synthesiser has been given in Chapter 4. In this section we will study how to control it.

Although Praat provides a graphic interface for the specification of the synthesiser's control parameters, all examples in this section will be given in the form of *Praat scripts*. A Praat script is a list of commands for Praat. The system comes with a useful and straightforward script editor for writing and running scripts and it is well documented in the user manual that is available via the program's Help menu.

In many ways, a Praat script could be regarded as a *score file* in the style of a Music N type of synthesis programming language, as introduced in Chapter 1. The main difference is that one cannot build instruments in Praat. One can, however, change certain characteristics of the synthesiser in order to create different speakers. It is possible to customise the size of the vocal tract, the thickness of the vocal folds, the size of the nasal cavity, and so on.

In order to synthesise a sound using the Praat physical model one needs to indicate: (a) the speaker and (b) a list of actions. In Praat parlance, the speaker is referred to as the *Speaker object* and the list of actions is referred to as the *Artwork object*. In practice, a script contains instructions for an *Artwork object* and a call for a *Speaker object* that should be defined beforehand in a separate file. For the sake of simplicity, the following examples will employ Praat's default adult female speaker model and we will refrain from changing its physical characteristics.

An *Artwork object* is defined in terms of breakpoint values (Figure 6.10) for 29 variables corresponding to the movements of 29 different components of the model, named after the corresponding muscles or parts of the human vocal system: *Interarytenoid, Cricothyroid, Vocalis, Thyroarytenoid, PosteriorCricoarytenoid, LateralCricoarytenoid, Stylohyoid, Sternohyoid, Thyropharyngeous, LowerConstrictor, MiddleConstrictor, UpperConstrictor, Sphincter, Hyoglossus, Styloglossus, Genioglossus, UpperTongue, LowerTongue, TransverseTongue, VerticalTongue, Risorius, OrbicularisOris, LevatorPalatini, TensorPalatini, Masseter, Mylohyoid, LateralPterygoid, Buccinator* and *Lungs*. These variables are highly dependent on one another, but for didactic reasons the following examples will be limited to a subset of 10 variables (Figure 6.11). In theory these variables can vary from −1.0 to +1.0, but in most cases we can assume that the starting point is at 0.0, which stands for the relaxed position of the corresponding part. The variables of our subset are:

- **Interarytenoid** (glottis): This variable influences the width of the glottis (i.e. the space between the vocal folds). It normally varies from 0.0 (vocal folds relaxed and open) to 1.0 (vocal folds stiffly closed). A value of 0.5 brings the vocal folds together into a position suitable for normal voicing.
- **Cricothyroid** (glottis): This variable influences the length and the tension of the vocal folds. It can normally vary from 0.0 to 1.0. Contraction of the cricothyroid muscle in a real vocal system causes a visor-like movement of the thyroid cartilage on the cricoid cartilage; this movement lengthens the vocal folds. The pitch of the utterance rises proportionally to the value of this variable.
- **LevatorPalatini** (velum): This variable controls the velopharyngeal port to the nasal cavities, where 0.0 opens the port and 1.0 closes it by lifting the velum. An open port (0.0) will nasalise the sound. Nasal sounds have fewer high-frequency components in their spectrum.
- **Genioglossus** (tongue): This variable corresponds to a muscle that pulls the back of the tongue towards the jawbone whilst the front of the tongue moves towards the hard palate. It normally ranges from 0.0 (resting position) to 1.0 (maximum pull).
- **Styloglossus** (tongue): This variable corresponds to a muscle that pulls the tongue body backwards and upwards. It normally varies from 0.0 (resting position) to 1.0 (maximum pull).
- **Hyoglossus** (tongue): This variable causes the tongue body to be pulled backwards and downwards. It normally varies from 0.0 (resting position) to 1.0 (maximally pulled).
- **Mylohyoid** (mouth): This variable influences the opening of the jaw and moves the back of the tongue towards the rear pharyngeal wall. It is normally set with values from 0.0 (resting position) to 1.0 (maximum jaw opening). A negative value (e.g. −0.5) can be used to raise the body of the tongue.

- **Masseter** (mouth): This variable influences the raising and lowering of the jaw. This parameter normally varies from −0.5 (jaw open) to 0.5 (jaw closed). The body of the tongue touches the palate when the jaw is closed.
- **OrbicularisOris** (mouth): This variable bears the name of a muscle that circles the lips. It influences the rounding and protrusion of the lips of the model and it normally varies between 0.0 (normal lips spread) to 1.0 (rounded with protrusion).
- **Lungs**: This variable is responsible for producing the airflow that sets the vocal folds into vibration during phonation. This variable can normally value between −0.5 and +1.5. The value −0.5 represents the maximum amount of air that can be exhaled by force and the value +1.5 represents the maximum amount of air that can be inhaled. A typical exhalation values 0.0 and a typical inhalation would not exceed 0.2 for normal voicing.

In the following examples, the remaining 19 variables of the model are set to the default system values. These default values normally correspond to the resting positions of the respective parts that they represent.

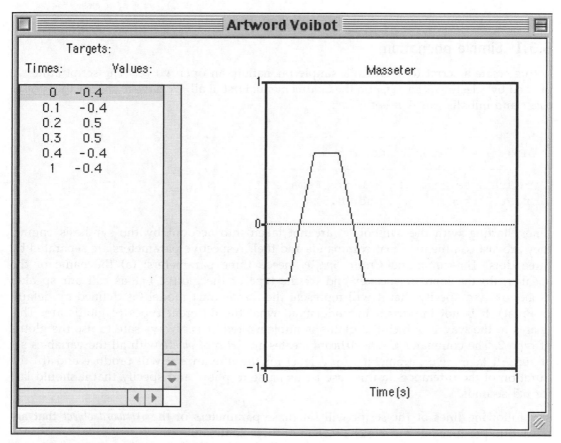

Figure 6.10 An example of breakpoint values for the *Masseter* variable. In this case the jaw is opened (−0.4) to start with, then it closes (0.5) for about 0.1 second and opens again

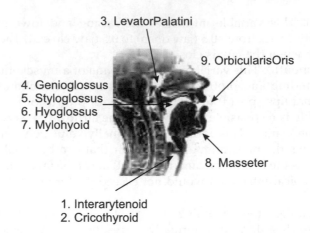

3. LevatorPalatini

9. OrbicularisOris

4. Genioglossus
5. Styloglossus
6. Hyoglossus
7. Mylohyoid

8. Masseter

1. Interarytenoid
2. Cricothyroid

Figure 6.11 The variables of the physical model that are used in the examples and their approximate counterparts of the human vocal system. (The lungs variable is not shown.)

6.3.1 Simple phonation

Let us create a script to produce a simple phonation: an open vowel. The script editor is opened by selecting *New script* on the *Control* menu. First of all, we have to specify the *Speaker object* and initialise an *Artword object*:

```
# -------------------------------------------
# Example 1: simple phonation
# -------------------------------------------
Create Speaker . . . Robovox Female 2
Create Artword . . . phon 0.5
```

Lines starting with the symbol '#' are not taken into account by the synthesis engine; they are just comments. Praat commands and their respective parameters are separated by three dots. The command *Create Speaker* needs three parameters: (a) the name of the speaker, (b) the kind of speaker and (c) the type of the glottis. Let us call our speaker 'Robovox' and specify that it will represent the *Female* tract model (as defined by default in Praat). It is not important to understand what the different types of glottis are. This relates to the way in which the glottis is implemented; it is always safe to use the glottis of type 2. The command *Create Artword* creates an *Artword object* with all the variables set at zero. It takes two parameters: (a) a label for the utterance it will produce and (b) the duration of the utterance. In this case let us name it 'phon' and specify that it should last for 0.5 seconds.

The following lines of the script will set those parameters of the *Artword object* that are needed to produce our vowel. The format of these commands is:

```
Set target . . . <time-point> <target-value> <synth-variable>
```

For example, the breakpoint values shown in Figure 6.10 are specified as follows:

```
Set target . . . 0.0  -0.4  Masseter
Set target . . . 0.1  -0.4  Masseter
Set target . . . 0.2   0.5  Masseter
Set target . . . 0.3   0.5  Masseter
Set target . . . 0.4  -0.4  Masseter
Set target . . . 1.0  -0.4  Masseter
```

In order to excite its vocal folds, Robovox needs to produce pressure flow by reducing the equilibrium width of the lungs. This is achieved by reducing the value of the *Lungs* variable from 0.1 to 0.0 during the initial 0.03 seconds of the utterance.

```
# ------------------------------------------
# Supply lung energy
# ------------------------------------------
Set target . . . 0.00  0.1  Lungs
Set target . . . 0.03  0.0  Lungs
Set target . . . 0.5   0.0  Lungs
```

Robovox has to keep its glottis closed with a certain stiffness so that the vocal folds can vibrate with the pressure coming from the lungs. This is achieved by setting the *Interarytenoid* and the *Crycothyroid* variables equal to 0.5 and 1.0, respectively, during the whole utterance, as follows:

```
# ------------------------------------------
# Control glottis
# ------------------------------------------
# Glottal closure
Set target . . . 0.0  0.5  Interarytenoid
Set target . . . 0.5  0.5  Interarytenoid
#
# Adduct vocal folds
Set target . . . 0.0  1.0  Cricothyroid
Set target . . . 0.5  1.0  Cricothyroid
```

Also, Robovox should close its velopharyngeal port to the nasal cavity in order not to let air scape through the nose, otherwise the sound will lack energy in its high-frequency components. Robovox's velopharyngeal port can be closed by setting the *LevatorPalatini* variable to 1.0 during the whole utterance:

```
# ------------------------------------------
# Close velopharyngeal port
# ------------------------------------------
Set target . . . 0.0  1.0  LevatorPalatini
Set target . . . 0.5  1.0  LevatorPalatini
```

Finally, Robovox should open its mouth and shape the vocal tract so as to produce a vowel. The mouth can be opened by setting the *Masseter* to –0.4. The *Hyoglossus* variable is set to 0.4

in order to pull Robovox's tongue backwards and downwards; this shapes the mouth to produce an open vowel.

```
# -------------------------------------------
# Shape mouth to open vowel
# -------------------------------------------
# Lower the jaw
Set target . . .  0.0  -0.4  Masseter
Set target . . .  0.5  -0.4  Masseter
#
# Pull tongue backwards
Set target . . .  0.0  0.4  Hyoglossus
Set target . . .  0.5  0.4  Hyoglossus
```

The next and final step is to activate the synthesis engine to produce the soundfile. This is done by selecting an *Artword object* plus a *Speaker object*, and by activating the synthesiser with the command *To Sound*.

```
# -------------------------------------------
# Synthesise the sound
# -------------------------------------------
Select Artword phon
plus Speaker Robovox
To Sound . . .  22050  25 0 0 0 0 0 0 0 0 0
```

The two first lines of the above code say that we wish to select the *Artword object* called 'phon' plus the *Speaker object* called 'Robovox'. The third line contains the command *To Sound*, whose first parameter is the sampling rate of the sound that will be synthesised (in this case 22 050 Hz) and the second is the oversampling coefficient. Oversampling is the number of times that the physical modelling equations will be computed during each sample period. The default value is 25; you should not need to modify this. It is not essential to understand the remaining nine parameters at this stage. They are used to generate charts reporting the behaviour of the model; we can simply leave these parameters as zeros. Finally, we select *Run* from the script editor menu in order to synthesise the sound. By default, Praat exhibits a window with a progress bar and an animation showing the behaviour of the model during the synthesis process. Once the synthesis process is completed, the sound appears in Praat's main window (it can be saved via the *Write* menu).

The complete script for this first example is given below and it will serve as the starting point for the examples that follow. This and other scripts (*example1.praat*, *example2.praat*, and so on) are available in the folder *robovox*, provided with the Praat system on the accompanying CD-ROM.

```
# -------------------------------------------
# Example 1: simple phonation
# -------------------------------------------
Create Speaker . . .  Robovox Female 2
Create Artword . . .  phon 0.5
```

```
# -----------------------------------------
# Supply lung energy
# -----------------------------------------
Set target . . .   0.00   0.1   Lungs
Set target . . .   0.03   0.0   Lungs
Set target . . .   0.5    0.0   Lungs
# -----------------------------------------
# Control glottis
# -----------------------------------------
# Glottal closure
Set target . . .   0.0   0.5   Interarytenoid
Set target . . .   0.5   0.5   Interarytenoid
#
# Adduct vocal folds
Set target . . .   0.0   1.0   Cricothyroid
Set target . . .   0.5   1.0   Cricothyroid
# -----------------------------------------
# Close velopharyngeal port
# -----------------------------------------
Set target . . .   0.0   1.0   LevatorPalatini
Set target . . .   0.5   1.0   LevatorPalatini
# -----------------------------------------
# Shape mouth to open vowel
# -----------------------------------------
# Lower the jaw
Set target . . .   0.0   -0.4   Masseter
Set target . . .   0.5   -0.4   Masseter
#
# Pull tongue backwards
Set target . . .   0.0   0.4   Hyoglossus
Set target . . .   0.5   0.4   Hyoglossus
# -----------------------------------------
# Synthesise the sound
# -----------------------------------------
Select Artword phon
plus Speaker Robovox
To Sound . . .   22050   25   0 0 0 0 0 0 0 0 0
```

Figure 6.12 portrays the first 0.2 second of the resulting sound (*phon1.aiff*; this and all other sound examples available in the folder *robovox* within the Praat materials on the CD-ROM). Note the highly dynamic nature of the very beginning of the signal, a phenomenon that confers a great degree of realism to the sound. This is very difficult to obtain with other synthesis methods. Figure 6.13 plots the excitation pressure at the bottom region of the trachea section of the model and Figure 6.14 shows the behaviour of the vocal folds in terms of the width of the glottis during vibration. After a rapid settling period, the glottis vibrates steadily in response to the regular pressure.

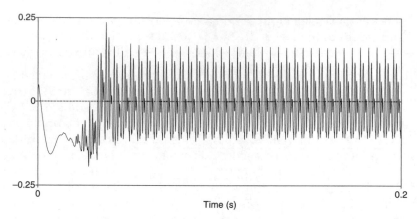

Figure 6.12 The first seconds of the simple phonation example. The highly dynamic nature of the beginning of the signal confers a great degree of realism to the sound

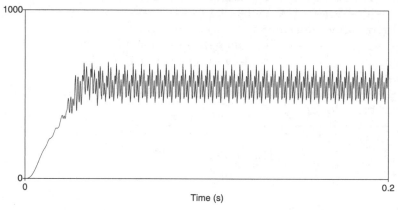

Figure 6.13 The excitation pressure that generated the simple phontation in Figure 6.12, measured at the bottom region of the trachea

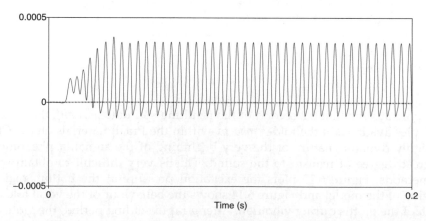

Figure 6.14 The width of the glottis during the production of the simple phonation in Figure 6.12

6.3.2 The effect of excessive lung pressure

In this example we will examine the effect of varying the *Lungs* variable from one extreme to the other: from 1.5 to –0.5. By replacing the following lines to the above script we obtain the sound displayed in Figure 6.15:

```
# -------------------------------
# Supply lung energy
# -------------------------------
Set target . . .  0.0   1.5   Lungs
Set target . . .  0.4  -0.5   Lungs
Set target . . .  0.5  -0.5   Lungs
```

The resulting sound (*phon2.aiff*) gets increasingly distorted as the lung pressure increases (Figure 6.16). Here we can clearly observe the effect of lung pressure on the amplitude of the

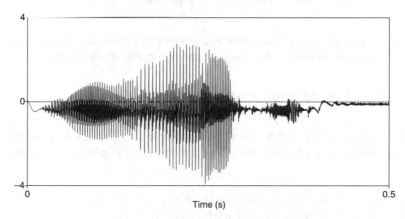

Figure 6.15 Sound generated under extreme lung pressure values. Too much pressure causes distortion of the sound

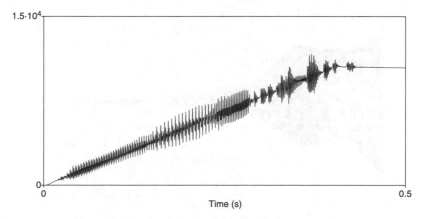

Figure 6.16 The excitation pressure that generated the sound in Figure 6.15 measured at the bottom region of the trachea

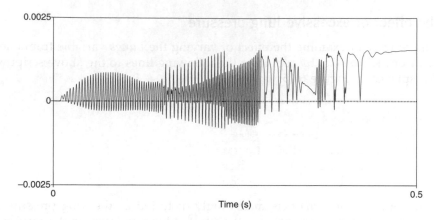

Figure 6.17 The width of the glottis during the production of the sound in Figure 6.15. Too much pressure gives rise to chaotic modes of vibration of the vocal folds

sound. Also, note in Figure 6.17 the damaging effect of excessive pressure on the glottis, whereby chaotic modes of vibration arise at approximately 0.3 seconds.

6.3.3 Pushing sustained phonation to its limits

The third example examines what happens if we instruct Robovox to sustain phonation for a long stretch of time (e.g. 13 seconds) without taking a breath. This is done by augmenting the duration of the *Artword object* in the script for our simple phonation example:

```
Create Artword . . . phon 13
```

In fact, all respective time-point values must be amended in this script; for example, the second breakpoint of the *Interarytenoid* variable should read as:

```
Set target . . . 13.0 0.5 Interarytenoid
```

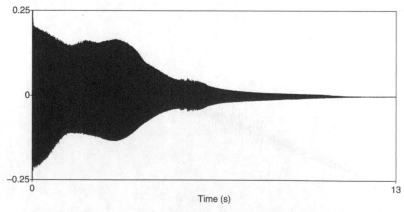

Figure 6.18 A 13-second phonation in one breath. During the first 5 seconds or so, the subglottal pressure falls steadily. As the pressure drops further, a chaotic period doubling stage suddenly takes place

The resulting sound is portrayed in Figure 6.18 (*phon3.aiff*) and the sub-glottal pressure registered in the bottom region of the trachea of the model is shown in Figure 6.19. During the first 5 seconds or so, the subglottal pressure falls steadily but then, a chaotic period doubling stage suddenly takes place. As the pressure drops even further, the register breaks, producing what could be regarded as a source of falsetto voice. Figure 6.20 shows what happens to the pitch of this sound. As a rule of thumb, pitch should fall as the pressure drops. However, at the period doubling stage the pitch displays unpredictable behaviour and jumps upwards.

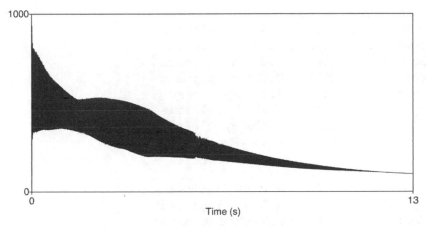

Figure 6.19 The excitation pressure that generated the sound in Figure 6.18 measured at the bottom region of the trachea

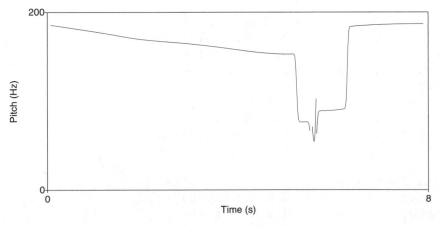

Figure 6.20 Pitch should normally fall as the pressure drops, but it can display unpredictable behaviour when the pressure is not suitable for normal phonation

6.3.4 Continuous vowel sequence

In this example, Robovox will produce a continuous sequence of vowels. This is achieved by adding new breakpoints to the *Artword* variables in order to continuously vary the shape of the vocal tract and the mouth during phonation. Let us set the duration of 'phon' to 4.0 and amend the following tongue and mouth breakpoint settings in the original simple phonation script:

```
# ------------------------------------------
# Tongue control
# ------------------------------------------
# Pull tongue back and downwards
Set target . . . 0.0     0.45   Hyoglossus
Set target . . . 0.95    0.45   Hyoglossus
Set target . . . 1.0     0.6    Hyoglossus
Set target . . . 2.3     0.6    Hyoglossus
Set target . . . 2.35   -0.2    Hyoglossus
Set target . . . 3.10   -0.2    Hyoglossus
Set target . . . 3.15    0.45   Hyoglossus
Set target . . . 4.0     0.45   Hyoglossus
#
# Pull tongue back and upwards
Set target . . . 0.0     0.0    Styloglossus
Set target . . . 0.95    0.0    Styloglossus
Set target . . . 1.0    -0.3    Styloglossus
Set target . . . 2.3    -0.3    Styloglossus
Set target . . . 2.35    0.2    Styloglossus
Set target . . . 3.10    0.2    Styloglossus
Set target . . . 3.15    0.0    Styloglossus
Set target . . . 4.0     0.0    Styloglossus
#
# Pull tongue towards jaw and palate
Set target . . . 0.0     0.0    Genioglossus
Set target . . . 0.95    0.0    Genioglossus
Set target . . . 1.0     0.0    Genioglossus
Set target . . . 2.3     0.0    Genioglossus
Set target . . . 2.35    0.5    Genioglossus
Set target . . . 3.10    0.5    Genioglossus
Set target . . . 3.15    0.0    Genioglossus
Set target . . . 4.0     0.0    Genioglossus
#
# Raises the tongue body
Set target . . . 0.0     0.0    Mylohyoid
Set target . . . 0.95    0.0    Mylohyoid
Set target . . . 1.0     0.0    Mylohyoid
Set target . . . 2.3     0.0    Mylohyoid
```

```
Set target . . .   2.35   -0.2   Mylohyoid
Set target . . .   3.10   -0.2   Mylohyoid
Set target . . .   3.15    0.0   Mylohyoid
Set target . . .   4.0     0.0   Mylohyoid
# ------------------------------------------
# Mouth control
# ------------------------------------------
# Lips rounding
Set target . . .   0.0     0.0   OrbicularisOris
Set target . . .   0.95    0.0   OrbicularisOris
Set target . . .   1.0     0.6   OrbicularisOris
Set target . . .   2.3     0.6   OrbicularisOris
Set target . . .   2.35   -0.2   OrbicularisOris
Set target . . .   3.10   -0.2   OrbicularisOris
Set target . . .   3.15    0.0   OrbicularisOris
Set target . . .   4.0     0.0   OrbicularisOris
#
# Jaw opening
Set target . . .   0.0    -0.45   Masseter
Set target . . .   0.95   -0.45   Masseter
Set target . . .   1.0    -0.2    Masseter
Set target . . .   2.3    -0.2    Masseter
Set target . . .   2.35   -0.16   Masseter
Set target . . .   3.10   -0.16   Masseter
Set target . . .   3.15   -0.45   Masseter
Set target . . .   4.0    -0.45   Masseter
```

Here Robovox will produce a sequence of four vowels (*phon4.aiff*). Figure 6.21 plots the pressure of the signal at the back of the mouth cavity. This well illustrates the effect of varying the shape of the vocal tract during phonation: different shapes result in the resonance and/or attenuation of different spectral components. Figure 6.22 shows the

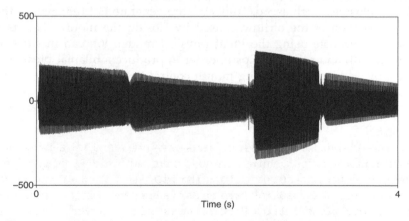

Figure 6.21 The pressure of the four-vowel signal at the back of the mouth cavity

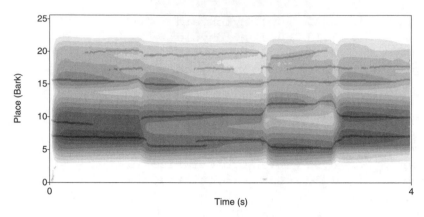

Figure 6.22 The cochleogram of the resulting four-vowel sequence

cochleogram of the resulting utterance; a cochleogram is an audiogram derived from electrophysiological prediction of hearing sensitivity, based on detection thresholds of the ear's cochlear responses to signals presented across a frequency range. Here we can clearly visualise the spectral differences between the four vowels.

6.3.5 Plosive bilabial consonants and beyond

The last example illustrates one of the features that makes this synthesiser so powerful: the ability to produce realistic consonants. Most vocal synthesisers attempt to produce consonants by means of a noise generator to synthesise a signal that resembles the turbulence produced by relaxed vocal folds. The standard practice is to simulate the activity of the vocal folds using two generators: one generator of white noise and one generator of a periodic harmonic pulse. Then, by carefully controlling the loudness of the signal produced by the generator, one can roughly simulate the production of syllables with consonants (refer to the example introduced earlier in section 6.2). The problem with this method is that it fails to produce realistic consonants that are not generated at the level of the vocal folds. For instance, bilabial plosives, such as /p/ and /b/, are generated at the level of the lips, as a result of the obstruction of the airflow caused by closing the mouth. This is reasonably straightforward to simulate using the Praat physical model. We can instruct Robovox to briefly close its mouth and round its lips in order to produce a bilabial plosive close to the consonant /b/, by adding the following mouth control to the script of the original simple phonation example:

```
# ---------------------------------------------
# Mouth control
# ---------------------------------------------
# Lips rounding
Set target . . . 0.0   0.0   OrbicularisOris
Set target . . . 0.1   0.0   OrbicularisOris
Set target . . . 0.2   1.0   OrbicularisOris
Set target . . . 0.3   1.0   OrbicularisOris
```

```
Set target . . .   0.4   0.0   OrbicularisOris
Set target . . .   0.5   0.0   OrbicularisOris
#
# Jaw opening
Set target . . .   0.0  -0.4   Masseter
Set target . . .   0.0  -0.4   Masseter
Set target . . .   0.2   0.5   Masseter
Set target . . .   0.3   0.5   Masseter
Set target . . .   0.4  -0.4   Masseter
Set target . . .   0.5  -0.4   Masseter
```

Figure 6.23 shows the resulting utterance. The interruption of the sound due to closing the mouth is very visible; this activity causes a plosive consonant effect. The highly sophisticated behaviour of the resulting sound during the closing and opening times is fascinating. This

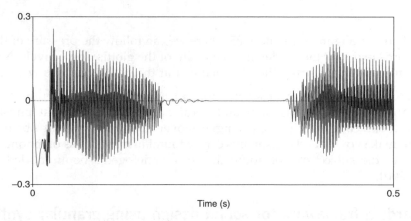

Figure 6.23 The utterance generated by briefly closing the mouth

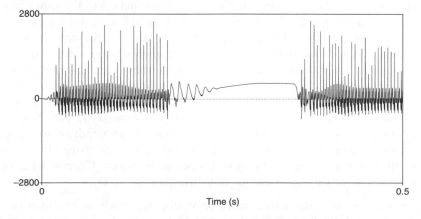

Figure 6.24 The excitation pressure that generated the sound in Figure 6.23 measured at the upper glottis region of the model

151

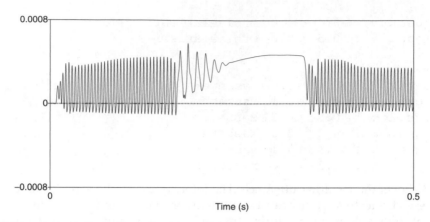

Figure 6.25 The width of the glottis during the production of the sound in Figure 6.23

can be seen better in Figures 6.24 and 6.25 where we can follow the pressure of the signal at the upper glottis region of the model and the width of the glottis, respectively. Note that the vocal folds stop vibrating during the obstruction, but the glottis remains wide open.

By careful control of airflow obstructions, tongue movements and mouth shape we can produce a great variety of consonant and vocal noises with the Praat physical model. Information about vocal tract shapes, tongue movements and types of obstructions that people use to make vowels and consonants can be found in most books on phonetics. A good introduction to the subject can be found in Peter Ladefoged's book entitled *Vowels and Consonants* (2001).

6.4 Towards a framework for sound design using granular synthesis

Chaosynth is a cellular automata-based granular synthesis system whose abilities to produce unusual complex dynamic sounds are limitless. (An introduction to the inner functioning of Chaosynth is given in section 5.1.4, Chapter 5. Also, refer to Chapter 8 for more information on the demonstration version of Chaosynth that is available on the accompanying the CD-ROM.) However, due to its newness and flexibility, potential users have found it very hard to explore its possibilities as there is no clear referential framework to grasp when designing sounds. Standard software synthesis systems take this framework for granted by adopting a taxonomy for synthesis instruments that has been inherited from the acoustic musical instruments tradition: i.e. woodwind, brass, string, percussion, etc. Sadly, the most interesting synthesised sounds that these standard systems can produce are simply referred to as *effects*. This scheme clearly does not meet the demands of more innovative software synthesisers. In order to resolve this issue, composer James Correa is developing an alternative taxonomy for Chaosynth timbres (Miranda *et al.*, 2000).

The development of the alternative taxonomy departs from a framework for sound categorisation inspired by Pierre Schaeffer's (1966) concept of sound maintenance (Figure 6.1) and to some extent by an article written by Jean-Claude Risset in the book *Le timbre:*

Table 6.1 Classes of Chaosynth sounds

Fixed mass	*Flow*	*Chaotic*	*Explosive*	*General textures*
lighten	cascade	insects	metallic	textures
darken	landing	melos	woody	gestures
dull	raising	boiler	glassy	
elastic	lift	windy	blower	
melted	crossing	noises	drum	
	drift			

métaphore pour la composition (1992). It is assumed that the most important characteristic of the sounds produced by Chaosynth is their spectral evolution in time.

The general structure of the proposed taxonomy is summarised in Table 6.1. The term 'instrument' is used below to refer to specific Chaosynth set-ups that produce the sounds of the respective categories. There are currently five general classes: *Fixed mass, Flow, Chaotic, Explosive* and *General textures*.

Sound example are available on the accompanying CD-ROM in the folder *taxonomy* within the Chaosynth materials.

6.4.1 Sounds of fixed mass

The *Fixed mass* class comprises those instruments that produce sounds formed by a large amount of small granules. The overall outcome of these instruments is perceived as sustained sounds with a high degree of internal redundancy; hence the label 'fixed mass'. Notice here that by fixed mass we do not mean fixed pitch, but rather a stable and steady spectrum where the frequencies of the grains are kept within a fixed band. Sometimes this creates a sense of pitch, but this phenomenon is not a mandatory condition. Correa has defined five sub-classes for this class: whilst *lighten* are bright instruments that produce sounds rich in high frequencies, *darken* are those instruments that produce sounds rich in low frequencies; *dull* instruments produce muffled sounds, and finally *elastic* and *melted* were named after the psychoacoustic sensation they create.

6.4.2 Flow instruments

These are instruments whose outcome gives a clear sensation of movement due to a continuous change of the fundamental frequencies of the grains. These sounds are said to have a variable mass, but not an unpredictable one as the frequencies of the granules tend to move collectively in one direction. Six sub-classes for this class were defined according to the direction of the spectral movement: *cascade* and *landing*, both comprise instruments whose outcome is in a descending direction, with the difference that the movement of the former is smooth and light, whereas the latter is fast and vigorous. Conversely, there are *raising* and *lift* instruments, where the movement is ascending. The *crossing* instruments produce sounds that display a cross-movement between two groups of frequencies, one

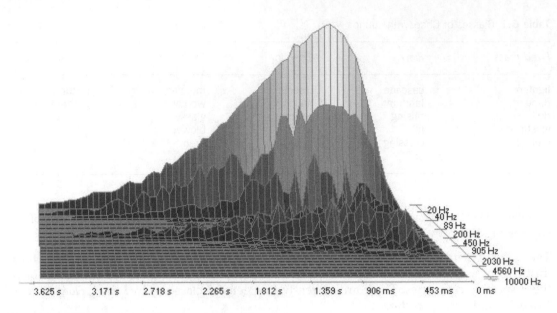

Figure 6.26 The spectrum of a cascade-type of sound

group going upwards and the other going downwards. Finally, the sub-class *drift* produces sounds whose sense of direction is not entirely clear; there is some flow but frequencies of the grains tend to drift up or down. Figure 6.26 shows the spectrogram of a *cascade-type* of sound (*cascade02.aiff* on the CD-ROM).

6.4.3 Chaotic instruments

This class of instruments produces sounds with a high degree of spectral movement. These sounds are of highly variable mass, the movement of the grain frequencies is unpredictable and they do not have an overall direction. We have defined five sub-classes of chaotic instruments: *insects*, are those instruments whose outcome resembles the noises produced by insects. These sounds have more activity in the high frequencies. *Melos* produce sounds whereby the individual grains can be perceived as individual notes as if they were fast melodic fragments, whereas the sounds from a *boiler* instrument resemble the sound of boiling liquids. *Windy* and *noises* produce sounds that resemble or are derived from white-noises: the former produces an auditory sensation of blowing air, whereas the latter produces noise similar to an AM radio off-station. Figure 6.27 portrays the spectrogram of a *melos* sound (*melos01.aiff* on the CD-ROM).

6.4.4 Explosive instruments

This class resembles traditional acoustic percussion instruments. The sounds here are of short duration, fast attack, practically no sustain and short decay; in general terms, these sounds have a short impulsion and are either of a fixed or variable mass. This class

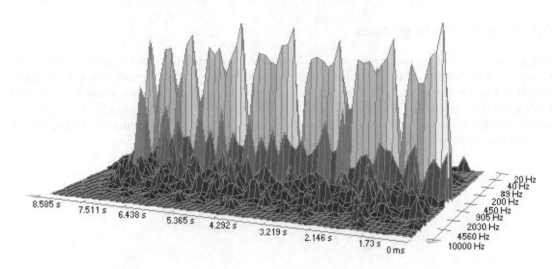

Figure 6.27 The spectrum of a melos sound

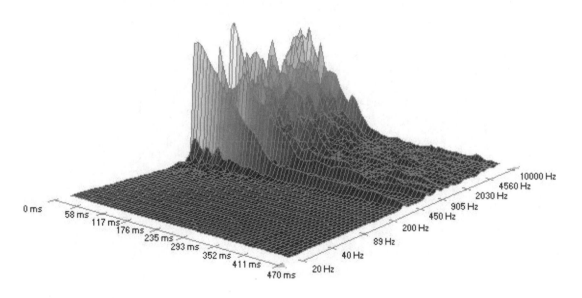

Figure 6.28 The spectrum of an explosive metallic sound

comprises five sub-classes, organised according to the resemblance of their sound to those from real percussive materials: *metallic, woody, glassy, blower* and *drum. Blower* instruments produce fast explosive-like sounds. In the *Drum* class we include those instruments whose sounds are very similar to membrane-type drums. Figure 6.28 shows the spectrum of a typical example of an *explosive metallic* sound (*metallic01.aiff* on the CD-ROM).

155

6.4.5 Textures and gestures

General textures are instruments that produce what we refer to *textures* and *gestures*. The outcome of *textures* is clouds of sounds with long duration and a variable mass. The classic example here is the stereotypical bubble and cloud-like texture that is normally associated with granular synthesis. The second sub-class, the *gestures*, generate sounds that resemble some standard musical instrument techniques, such as a cello glissando, or metaphorical sounds such as the noise of a steam train, a helicopter or a Klingon phaser.

7 Towards the cutting edge: artificial intelligence, supercomputing and evolutionary systems

7.1 Artificial intelligence sound synthesis

The sound domain of Western music is no longer demarcated by the boundaries of traditional acoustic instruments. Nowadays, composers have the opportunity to create music with an infinite variety of sounds, ranging from those produced by acoustic devices and different types of mechanical excitation to electronically synthesised sounds.

It is evident from the previous chapters that computer technology offers the most detailed control of the internal parameters of synthesised sounds, which enables composers to become more ambitious in their quest for more sophisticated and expressive electronic sonorities. In this case, however, the task of sound synthesis becomes more complex. A composer can set the parameters for the production of an immeasurable variety of sounds, but this is often accomplished by feeding a synthesis algorithm with streams of numerical data, either by means of an external controller (e.g. a MIDI controller) or via a score (e.g. as in pcmusic). Even if the composer knows the role played by each single parameter for synthesising a sound, it is both very difficult and tedious to ascertain which values will synthesise the sound he or she wants to produce; the previous chapter's physical modelling vocal synthesis exercise is a good example of this. In such a situation, higher processes of inventive creativity and abstraction often become subsidiary to time-consuming, non-musical tasks.

Recent studies in acoustics, psychoacoustics and the psychology of music have vastly expanded our knowledge of the nature and perception of sounds and music. It seems, however, that this interdisciplinary knowledge we have about the nature and perception of sounds has not yet been taken into account by sound design software technology. Better systems may certainly be provided if engineers devise ways of including this knowledge in a sound synthesis software. One approach to this challenge is to combine sound synthesis engineering with artificial intelligence (AI) techniques.

The first section of this chapter introduces ongoing research work being conducted by the author. We are investigating methods to combine synthesis engineering with AI techniques in order to provide sound synthesis systems with the interdisciplinary knowledge mentioned above. In the quest for more user-friendly sound synthesis programs we have designed a case study system called Artist (short for Artificial Intelligence Synthesis Tool). Artist takes a radically different route from the majority of programs currently available. Whilst most programs rely on the use of sophisticated graphic interfacing to facilitate they operation, Artist concentrates on the use of natural language; for example, the user may type or speak commands such as 'play a sound with high acuteness' or 'play a dull low-pitched sound' (Miranda, 1995b).

We do not claim that Artist's operational method is a replacement for graphic interfacing. On the contrary, we believe that both methods are complementary. Nevertheless, Artist's approach may be more suitable for a number of other applications where graphic interfacing alone does not suffice; for example, for people with impaired visual, neurological or muscular abilities.

7.1.1 The quest for artificial intelligence

One of the main objectives of AI is to gain a better understanding of intelligence but not necessarily by studying the inner functioning of the brain. The methodology of AI research is largely based upon logic, mathematical models and computer simulations of intelligent behaviour.

Of the many disciplines engaged in gaining a better understanding of intelligence, AI is one of the few that has particular interest in testing its hypotheses in practical day-to-day situations. The obvious benefit of this aspect of AI is the development of technology to make machines simulate diverse types of intelligent behaviour. For example, as a result of AI development, computers can play chess and diagnose certain types of diseases extremely well.

It is generally stated that AI as such was 'born' in the late 1940s, when mathematicians began to investigate whether it would be possible to solve complex logical problems by performing sequences of simple logical operations automatically. In fact, AI may be traced back to far before computers were available, when mechanical devices began to perform tasks previously performed only by the human mind. For example, in the 1830s, Charles Babbage, an English mathematician, invented the *Analytical Engine*, a mechanical machine that could automatically execute complex sequences of symbolic operations.

7.1.2 Introduction to Artist

Designing sounds on a computer is certainly a complex kind of intelligent behaviour. Composers engage in cognitive and physical acts in order to establish the suitability and effectiveness of their creations prior to constructing them. In attempting to solve a sound design problem, composers explore possible solutions by trying out possibilities and investigating their consequences.

When synthesising sounds to be used in a composition, composers generally have their own ideas about the possibilities of organising these sounds into a musical structure. In attempting to obtain the desired sound, the composer explores a variety of possible solutions, trying out those possibilities within his or her personal aesthetic. It is this process of exploration that has instigated our research work. This process frequently results in inconsistencies between the composer's best guess at a solution and the formulation of the problem. If no solution is found then the problem either has no solution at all or it must be redefined. Sound synthesis is seen in this context as a task which demands, on the one hand, clarification of the problem and, on the other, the provision of alternative solutions. As an example, envisage a situation in which a musician commands Artist to produce a high-pitched sound. In order to do this the system might need the expression 'high-pitched sound' to be clarified. In this case the musician would explain that 'high-pitched' means a fundamental frequency above a certain threshold. If the system still does not understand the clarification, then at least some sound should be produced (e.g. at random), which would give some chance that the sound produced might satisfy the musician's requirement. This sound may then be taken as a starting point for refinement; e.g. the musician could command the computer to 'transpose it to a high-frequency band'.

In order to aid this process, Artist focuses on the provision of the following features:

- The ability to operate the system by means of an intuitive vocabulary instead of sound synthesis numerical values. It must be said, however, that the coherence of this vocabulary depends upon several factors, ranging from how the user sets up the synthesis algorithms to the terms (or labels) of the vocabulary of sound descriptors
- The ability to customise the system according to particular needs, ranging from defining which synthesis technique will be used to the vocabulary for communication (for example, verbs and adjectives in English)
- The encouragement of the use of the computer as a collaborator in the process of exploring ideas
- The ability to aid the user in concept formation, such as the generalisation of common characteristics among sounds and their classification according to prominent sound attributes
- The ability to create contexts which exhibit some degree of uncertainty

Currently Artist has two levels of operation: *instrument design* and *sound design*. At the instrument design level, the user customises the system by implementing a synthesis algorithm and specifying the basic terms of a vocabulary (this vocabulary may grow automatically during sound design operations). At the sound design level, the system provides facilities for the design of sounds with the synthesis algorithm and vocabulary to hand.

We suggest a method for the instrument design level which assumes that the vocabulary for sound description is intimately related to the architecture of the instrument. If one wishes to manipulate the vibrato quality of sounds, for example, it is obvious that the vibrato generator must be available for operation. As far as the sound design level is concerned, the user is required to know a few basic Artist operational commands, the capabilities of the instrument at hand and the available vocabulary for sound description.

7.1.3 Representing and processing knowledge of sound synthesis

Artist requires relatively sophisticated means to store and process data. It is not only a matter of storing synthesis parameter values in a traditional database; Artist must also store knowledge of sound synthesis as well as knowledge of the system itself (for example, the structure of the instrument, the role of each synthesis parameter and the meaning of the vocabulary in terms of synthesis parameter values, to cite but a few).

We designed the storage methods and processing mechanisms of Artist based upon a number of hypotheses on how the human mind would store and process knowledge of sound synthesis. We introduce below a selection from our hypotheses and their respective roles in Artist. Note that these hypotheses are rather pragmatic and are by no means intended to explain how the human mind works.

Hypothesis 1: the layered organisation of knowledge

Our first hypothesis is that humans tend to use a layered approach to the organisation of knowledge – including knowledge of sounds. We believe that when people think of a sound event they tend to identify its perceptual qualities and regard them as kinds of assembled perceptual units. Together, these perceptual units form a concept in the mind. This concept is part of our knowledge of the world of sound and it is connected to other concepts of the same domain through appropriate relationships.

In consequence of this hypothesis, we believe that people tend to represent information about sounds in a layered manner and use the higher-level layers to recall them. Taking as an example the sound of thunder: one might associate it with attributes such as loud amplitude, sharp attack, noisy, low pitch, and medium duration. In this case, we would say that people tend to group this information and recall it simply as *thunder* instead of as *loud–sharp–noisy–low–short* (Figure 7.1).

In order to program this hypothesis we devised a representation technique called *Abstract Sound Schema* (ASS). The ASS is a tree-like abstract structure consisting of *nodes*, *slots* and *links*. Nodes and slots are components and links correspond to the relations between them. Both components and relations are labelled on the ASS. Slots are grouped bottom-up into higher-level nodes, which in turn are grouped into higher-level nodes and so on until the top node (Figure 7.2).

The role of the ASS is twofold: it embodies a multi-level representation of the synthesis instrument and also provides an abstraction to represent sounds (Figure 7.3). In practice, each slot has a user-defined label and accommodates either a sound synthesis datum or a pointer towards a procedure to calculate this datum. Higher-level nodes also have user-

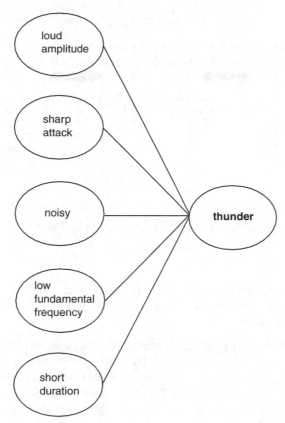

Figure 7.1 Humans tend to use a layered approach to the organisation of knowledge. When we think of a sound event, we tend to identify its perceptual qualities and regard them as kinds of assembled perceptual units

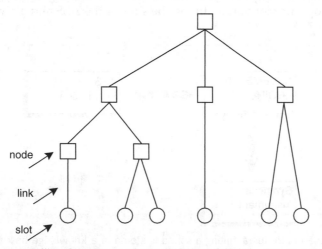

Figure 7.2 The ASS is a tree-like abstract structure consisting of nodes, slots and links. Nodes and slots are components and links correspond to the relations between them

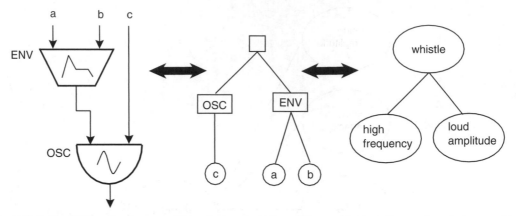

Figure 7.3 The ASS embodies a multi-level representation of the synthesis instrument and also provides an abstraction to represent sounds

defined labels and they correspond to the modules and sub-modules of the synthesis instrument. The top node – the root of the tree, in AI jargon – is therefore an abstraction that accommodates a sound event.

The slots must be loaded with synthesis parameter values in order to instantiate a certain sound. For each different sound produced by the instrument there is a corresponding instantiation. For example, considering the case of a subtractive synthesis instrument which simulates the vocal tract mechanism (see Chapter 6) an instantiation would correspond to the set of parameters for the simulation of a specific position of the vocal tract when producing a certain sound.

The synthesis values for instantiating sounds are recorded in a knowledge base as a collection of slot values. To produce a sound, an *assembler engine* is then responsible for collecting the appropriate slot values, filling the slots of the ASS and activating the synthesis instrument (Figure 7.4).

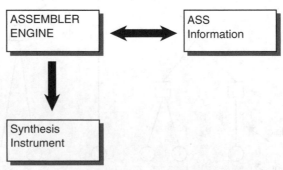

Figure 7.4 ASS information for instantiating a sound is stored in a knowledge base as a collection of slot values. An assembler engine is responsible for collecting the appropriate slot values in order to activate the synthesis instrument

Hypothesis 2: the identification of sound kinship

Our second hypothesis is that, in most cases, sound kinship can be defined within a multi-dimensional *attribute space*. This hypothesis is influenced by several experiments by David Ehresman and David Wessel (1978) and David Wessel (1979).

The basic idea is illustrated in Figure 7.5. If, for example, we form a three-dimensional space from three attributes, so that each attribute varies continuously; given sounds A, B and C, whose three attributes correspond to coordinates of this space, there is a sound D, such that A, B, C and D constitute the vertices of a parallelogram. In this context, sound D is expected to be perceptually akin to sounds A, B and C. To illustrate this, imagine a configuration in which sounds are described by three attributes corresponding to the axes *x*, *y* and *z*: first formant centre frequency (represented as F1), second formant centre frequency (F2) and fundamental frequency (F0), respectively. On the one hand, *sound(a)* and *sound(b)* may have similar values for axes *x* and *y*, that is, F1 = 290 Hz and F2 = 1870 Hz, whilst, on the other hand, *sound(a)* and *sound(c)* have the same value for axis *z*, namely F0 = 220 Hz, and the value of F0 for *sound(b)* is equal to 440 Hz. If one makes a two-sound pattern – *sound(a)* followed *sound(b)* – and wishes to make an analogous pattern beginning on *sound(c)*, then according to our hypothesis the best choice would be the sound whose attributes best complete a parallelogram in the space, that is, *sound(d)*. In perceptual terms, what has varied is the attribute pitch: *sound(b)* sounds higher than *sound(a)* but preserved the same formant

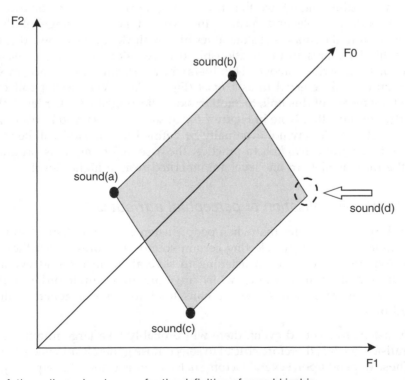

Figure 7.5 A three-dimensional space for the definition of sound kinship

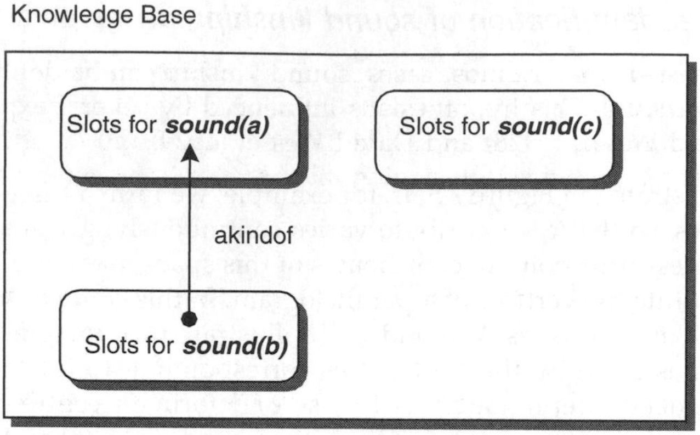

Knowledge Base

Figure 7.6 The link *akindof* is a structural link that allows for the specification of inheritance relationships in the knowledge base

structure. Therefore in order to obtain an analogous two-sound pattern beginning on *sound(c)* (say, F1 = 650 Hz and F2 = 1028 Hz), the best solution would be *sound(d)*, with F1 and F2 inherited from *sound(c)* but F0 inherited from *sound(b)*.

Although the kinship hypothesis may work well only for optimal cases (for example, it might not suit larger dimensions), we find it encouraging that a systematic description of sound events is both possible and meaningful. Artist takes advantage of the kinship hypothesis by recording the clusters of slot values hierarchically in the knowledge base. This ability to hierarchically represent the relationship between slot collections is useful for the *inheritance mechanism* described above. This hierarchical organisation is accomplished by means of a structural link referred to as *akindof* (Figure 7.6). When a slot collection for a sound is related, by means of this link, to another slot collection at a higher level, the former inherits properties of the latter. Note in Figure 7.6 that *sound(b)* is said to be *akindof sound(a)*. This means that slots which may not eventually be defined for *sound(b)* will be instantiated with slot values taken from *sound(a)*. In practice, the assembler engine is programmed to consider that the missing slots in one level are inherited from a higher level.

Hypothesis 3: the generalisation of perceptual attributes

Finally, our third hypothesis states that when people listen to several distinct sound events, they tend to characterise them by selecting certain sound attributes which they think are important. We hypothesise that when listening to several distinct sound events, people prioritise the selection of certain attributes which are more important in order to distinguish between them. We say that in this case humans tend to make a generalisation of the perceptual attributes.

If one carefully listens to a sound event, there will probably be a large number of possible intuitive generalisations. It is therefore crucial to select those generalisations we believe to be appropriate. These depend upon several factors such as context, sound complexity, duration of events, sequence of exposure and repetition, which make a great variety of combinations possible. Humans are, however, able to make generalisations very quickly; perhaps because we never evaluate all the possibilities. We tend to limit our field of exploration and resort to a certain evaluation method. We believe that this plays an important role in imagination and memory when designing sounds and composing with them.

Artist uses inductive machine-learning techniques in order to forge this third hypothesis. These techniques are widely used in AI systems to induce general concept descriptions of objects from a set of examples (Luger and Stubblefield, 1989). In Artist we use these techniques to infer which attributes are more distinctive in a sound. The term 'more distinctive' in this case does not necessarily refer to what humans would perceive to be the most distinctive attribute of a sound. Current machine-learning techniques are not yet able to mimic all the types of strategies used by humans. Nevertheless, we propose that certain kinds of human strategies might use information theory to make generalisations; these type of strategies can be programmed on a computer.

Artist therefore employs machine-learning algorithms that use information theory to automatically 'induce' which attributes are more distinctive on sounds given in a training set (Miranda, 1997). In this case, the training set is normally gathered by Artist itself from its own knowledge base; that is, from its collection of slot values. The result of learning is the description of sounds (or classes of sounds) in the form of rules. For example, a rule for a type of sound called *open vowel* could be:

open vowel: vibrato = normal,
formant resonators = vowel/a/
gender = male

or

vibrato = low
formant resonators = vowel/e/
gender = female

The interpretation of the above rule is: a sound event is an open vowel if it has normal vibrato rate, its spectral envelope corresponds to the resonance of a vowel/a/, and it is a male-like vocal sound; or it has a low vibrato rate, its spectral envelope corresponds to the resonance of a vowel/e/, and it is a female-like vocal sound.

The aim of inducing such rules is to allow the user to explore further alternatives when designing particular sounds. The user could ask Artist, for example, to 'play something that sounds similar to an open vowel'. In this case Artist would infer from the above rule which attributes are imperative to synthesise this type of sound. Before we present an example operation and study Artist's machine-learning module, let us examine how the system supports high-level symbolic sound description.

7.1.4 Describing sounds by means of their attributes

It is important to emphasise that we distinguish between timbre and spectrum. The latter characterises the physical structure of a sound and it normally uses mathematics and scientific jargon for description. The former denotes the perception of these properties, and the vocabulary for its description can be highly subjective. The term 'attribute' is used in the context of Artist to refer to perceived characteristics of the timbre of a sound.

The ASS provides a good framework for sound description, but it does not, however, prescribe a methodology for describing sounds coherently. Much research is still needed on this front. We identified a variety of approaches systematically to describe sounds from their attributes (Bismark, 1974; Terhardt, 1974; Schaeffer, 1966; Slawson, 1985). It is beyond the scope of this book to survey all of these, so we have selected one approach for discussion: the *functional approach*. Here one specifies a vocabulary of sound descriptors by using the architecture of the instrument as a point of departure. The expression 'vocabulary of sound descriptors' refers to a set of user-defined labels that are used to denote a combination, or group, of synthesis parameter values. Note that the ASS is well suited for the functional approach.

To illustrate the functional approach to sound description, consider the subtractive formant synthesiser in Figure 7.7; the reader is invited to refer to Chapter 6 for a detailed description of this instrument. The voicing source model of this instrument is depicted in Figure 7.8 and the ASS representation of this module is illustrated in Figure 7.9. Here one could create an

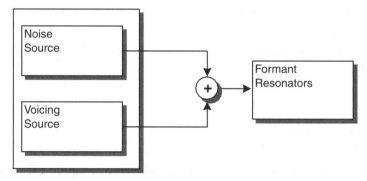

Figure 7.7 The modules of a subtractive synthesis instrument

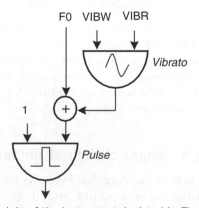

Figure 7.8 The voicing source module of the instrument depicted in Figure 7.7

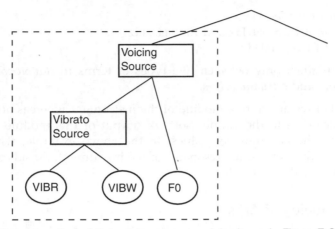

Figure 7.9 The ASS representation of the voicing source module shown in Figure 7.8

attribute labelled as 'voicing source' and ascribe descriptive values to it according to certain rules. For instance: *voicing source = steady low* if F0 = 80 Hz, VIBR = 5.2 Hz and VIBW = 3.5 per cent of F0 (VIBR and VIBW stand for vibrato rate and vibrato width, respectively).

Taking another example, let us attach an attribute labelled as 'openness' to the first formant resonator of the instrument; assume that the resonator module of the instrument portrayed in Figure 7.7 is a bank of band-pass filters (Figure 7.10). In this case, the higher the value of the frequency of the first formant, the greater the openness of the sound. For example, one could define three perceptual stages, each corresponding to a certain first formant centre frequency range of values:

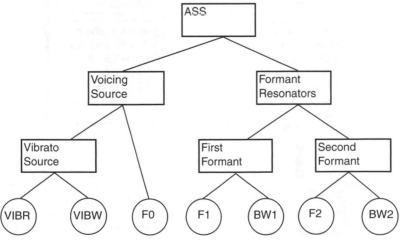

Figure 7.10 The right branch of the ASS represents the formant resonator modules of the instrument portrayed in Figure 7.7

- *openness* = *low* if F1 =< 380 Hz
- *openness* = *medium* if F1 > 380 Hz and F1 < 600 Hz
- *openness* = *high* if F1 >= 600 Hz

Note that we distinguish only between two types of terms in our vocabulary for sound description: *attribute* and *attribute values*.

Attribute is the label we attach to a module of the instrument whereas attribute values are the labels which we give to the various sorts of output of this module, according to the variations in its synthesis parameter values. In the above example, whilst the attribute openness is given to the first format resonator of the instrument, the adjectives *low*, *medium* and *high* are values for this attribute.

7.1.5 The functioning of Artist

To summarise, a sound event in Artist is composed of several sound attributes. Each sound attribute in turn corresponds to one or a set of synthesis parameters. An assembler engine is responsible for computing the necessary synthesis parameter values in order to produce a sound event. The input for the assembler engine can be either the name of the desired sound event, or a set of attribute values; alternatively, one can also input numerical synthesis parameter values. The output from the assembler engine is a set of synthesis parameter values which in turn are used for synthesising the required sound event (Figure 7.11).

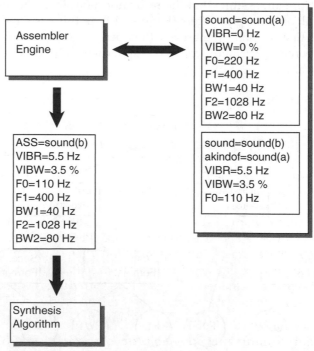

Figure 7.11 The functioning of Artist. The assembler engine deduces the synthesis parameters values for synthesising a sound event

The assembler engine computes all the necessary synthesis parameter values for synthesising a sound event. Since the input requirement is often ill-defined (for example, an incomplete list of sound attribute values), the assembler engine is programmed to deduce the missing information by making analogies with other sound events that have similar constituents. To do this, the system automatically induces rules of prominent sound features that will identify which sound attributes are more important for describing sound events. Furthermore, given that synthesis parameter values may also be given as part of the requirement, the system also handles values which may not match any available sound attribute. Likewise, the system automatically adds this new information to the knowledge base.

As an example, suppose that Artist has the following information in its knowledge:

- Sound events = { sound(a), sound(b) }
- Sound attributes = { openness = { low, high }, acuteness = { low, high }, fundamental frequency = { low, medium, high } }
- Synthesis parameters = { F1 = { 290 Hz, 400 Hz, 650 Hz }, F2 = {1028 Hz, 1700 Hz, 1870 Hz}, F0 = {220 Hz, 440 Hz, 880 Hz} }

The symbols in brackets denote possible values for the element on the left-hand side of the equation. For example, a sound attribute may value *openness*, *acuteness*, or *fundamental frequency*. The sound attribute *openness* may in turn value either *low* or *high*.

The assembler engine is programmed to infer from the system's ASS that, for example, *sound(a)* is described by having three attributes and values: *low openness*, *high acuteness* and *low fundamental frequency*. Furthermore, it can infer that *low openness* and *high acuteness* mean F1 = 290 Hz and F2 = 1870 Hz, respectively and that *low fundamental frequency* means F0 = 220 Hz.

Suppose that one inputs a request to synthesise a sound but only the information *low fundamental frequency* is given. In this case, Artist considers this input as an ill-defined requirement because the two other attributes needed to describe a sound fully are missing; in other words, there are no values for openness and acuteness. In this case, the assembler engine must allocate values for the missing attributes. This is accomplished by either consulting the induced rules of prominent sound features (the ones that identify which sound attributes are more relevant for describing a certain class of sound events) or by allocating values by chance. For this example, let us suppose that Artist synthesises *sound(a)* because it knows some rules suggesting that this sound best matches the description *low fundamental frequency*.

7.1.6 Machine generalisation of perceptual attributes

Machine learning (ML) is a major subfield of AI research with its own various branches (Carbonell, 1990). Perhaps the most popular current debate in machine learning, and in AI in general, is between the subsymbolic and the symbolic approaches. The former, also known as connectionism, or neural networks, is inspired by neurophysiology; its proponents intend to provide mechanisms so that the desired computation can be achieved by repeatedly exposing a system to an example of the desired behaviour. As the result of learning, the system records the behaviour in a network of single processors metaphorically called 'neurones'.

Artist currently uses the symbolic approach. Several algorithms for symbolic learning have been employed in AI systems. These range from learning by being told, to learning by discovery. In the former case, the learner is told explicitly what is to be taught by a 'teacher'. In learning by discovery, the system automatically discovers new concepts, merely from observations, or by planning and performing experiments in the domain. Many other techniques lie between these two extremes. The criterion for selecting a learning technique depends upon many factors, including its purpose and the representation scheme at hand.

The purpose of ML in Artist is to induce general concept descriptions of sounds from a set of examples. The ML technique selected for our investigation is therefore the *inductive learning technique* (IML). The benefit of being able to induce general concept descriptions of sounds is that the machine can automatically use induced concept descriptions to identify unknown sounds and to suggest missing attributes of an incomplete sound description.

IML provides Artist with the ability to make generalisations in order to infer which attributes are 'more distinctive' in a sound. The term 'more distinctive' in this case does not necessarily refer to what humans would perceive to be the most audibly distinctive attribute of a sound. The IML algorithm used in Artist uses information theory to pursue this task. As mentioned earlier, current ML techniques are not yet able to mimic all the types of heuristics used by humans. Nevertheless, it is proposed that one kind of heuristic might use information theory to make generalisations.

Once the generalisations have been made, the user may employ the descriptive rules to specify new sounds, different from those that were originally picked out as typical of the sounds that the system already knows.

The aim of inducing rules about sounds is to allow the user to explore further alternatives when designing particular sounds. The user could ask the system, for example, to 'play something that sounds similar to a vowel' or even 'play a kind of dull, low-pitched sound'. In these cases the system would consult induced rules to infer which attributes would be necessary to synthesise a vowel-like sound, or a sound with a dull colour attribute.

The inductive machine learning engine

The target of IML in Artist is to induce concepts about sounds represented in a training set. The training set can be either new data input by the user, or automatically inferred by the system from its own knowledge base.

Inductive learning can be either *incremental*, modifying its concepts in response to each training example, or *single trial*, forming concepts once in response to all data. A classic example of incremental inductive learning is a program called Arches (Winston, 1985). Arches is able to learn the structural description of an arch from examples and counter-examples supplied by a 'teacher'. The examples are processed sequentially and Arches gradually updates its current definition of the concept being learned by enhancing either the generality or the specificity of the description of an arch. It enhances the generality of the description in order to make the description match a given positive example, or the specificity in order to prevent the description from matching a counter-example.

The Iterative Dichotomizer 3 (ID3) algorithm is a classic example of single trial inductive learning (Quinlan, 1986). ID3 induces a decision tree from a set of examples of objects of a domain; the tree classifies these objects according to their attributes. Each example of the training set is described by a number of attributes. The ID3 algorithm builds a decision tree by measuring all the attributes in terms of their effectiveness in partitioning the set of target classes; the best attribute (from a statistical standpoint) is then elected as the root of the tree and each branch corresponds to a partition of the classifications (i.e. values of this attribute). The algorithm recurs on each branch in order to process the remaining attributes, until all branches lead to single classification leaves.

In principle, any technique that produces classificatory rules based upon feature descriptions could be useful for Artist. Ideally the system should use various IML algorithms in order to provide more than one classificatory possibility. The ability to have more than one classificatory possibility is useful in a situation where, for example, the user inputs a request to produce a sound and the system must check whether it knows a sound matching this request. Therefore by having more than one classificatory rule, the system has a greater chance of finding a matching sound and indeed of finding more than one sound which satisfies the requirement. To this end, Artist currently uses two single trial IML algorithms: the Induction of the Shortest Concept Description (ISCD) and the Induction of Decision Trees (IDT) (Dietterich and Michalski, 1981; Bratko, 1990).

The ISCD algorithm aims to induce the shortest description(s), that is, the smallest set(s) of attribute values of a sound (or class of sounds) which can differentiate it from the others in the training set. The IDT algorithm also induces classificatory rules, but not necessarily the most succinct ones. This book will focus only on Artist's IDT algorithm.

Artist's IDT algorithm is an adapted version of a Quinlan-like algorithm described in (Bratko, 1990). The result is represented in the form of a Decision Tree (DT), where internal nodes are labelled with attributes, and branches are labelled with attribute values (note, however, that the DT is not the same as the ASS representation discussed earlier). The leaves of the tree are labelled with sound classes. To classify a sound event, a path in the tree is traversed, starting at the root node and ending at a leaf. The IDT algorithm (refer to Appendix 3) proceeds by searching at each non-terminal node for the attribute whose values provide the best discrimination among the other attributes, that is, the Most Informative Attribute (MIA); the formula for the selection of the MIA is explained in Miranda, 1994.

Figure 7.12 shows an example DT induced from an hypothetical training set as follows:

Sound Name: *dull*
  Sound Attributes:
    openness = *narrow*
    pitch = *high*
    vibrato rate = *default*
  Sound Name: *wobbly*
  Sound Attributes:
    openness = *wide*
    pitch = *low*
    vibrato rate = *fast*

etc.

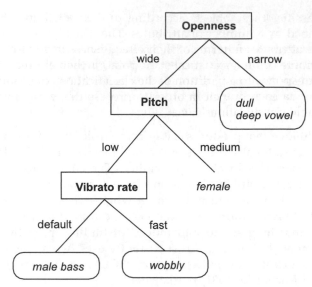

Figure 7.12 An example DT induced from an hypothetical training set

Once the decision tree is induced, a path is traversed in the tree to identify a sound, starting at the root (top sound attribute) and ending at a leaf. One follows the branch labelled by the attribute value at each internal node. For example, a sound described as having 'wide openness, low pitch and fast vibrato rate' is classified, according to this tree, as *wobbly*.

Artist is not included on the accompanying CD-ROM because it exists only as an experimental prototype programmed in Prolog, a programming language for implementing Artificial Intelligence experiments.

7.2 Supercomputing and sound synthesis

Since computers were invented there have been no major changes in their basic design. The main core of most currently available computers is a central processing unit (CPU), provided with a single memory component, which executes instructions sequentially. This configuration is commonly referred to as *von Neumann architecture* (after its inventor), but for the purposes of this book we use the more generic acronym SISD, for Single Instruction Single Data.

SISD machines call for the concept of sequential programming in which a program consists of a list of instructions to be executed one after the other. Although such machines may now work so fast that they appear to be executing more than one instruction at a time, the sequential programming paradigm still remains for most techniques and tools for software development, including sound synthesis.

The sequential paradigm works well for most of the ordinary tasks for which we need the computer but scientists are very aware of the limitations of current computer technology. For example, if computers were programmed to display intelligent or customary behaviour the sequential paradigm would fail to model many aspects of human intelligence and natural systems. Particular mental processes seem to be better modelled as a system distributed over thousands of processing units as, for example, an idealised model of brain tissue. It is possible to simulate this distributed type of behaviour on fast SISD machines by using a limited number of processing units, but no processor can currently realistically support the number of units required to study the behaviour of the more sophisticated distributed machine models.

Although the speed of SISD machines has increased substantially over the last few years, these improvements are beginning to reach the limits of physics, beyond which any further development would be impossible. In particular, the speed of light is a constant upper boundary for the speed of signals within a computer, and this speed is already proving too slow to allow further increases in the speed of processors. It seems that the only possible way to meet the demand for increasing performance is to abandon dependence upon single-processor hardware and look towards parallel processing, or *parallel supercomputing*.

As far as sound synthesis is concerned, parallel supercomputing is commonly associated with the possibility of breaking the speed limits imposed by SISD machines. However much faster sequential computers become, parallel computing will still offer interesting programming paradigms for the development of new synthesis techniques. From a composer's perspective, it is this latter aspect of parallel computing that excites interest.

7.2.1 Architectures for parallel supercomputing

Two main approaches are used to build parallel supercomputers: SIMD (for Single Instruction Multiple Data) and MIMD (for Multiple Instructions Multiple Data). Whilst the former employs many processors simultaneously to execute the same program on different data, the latter employs several processors to execute different instructions on different data. Both approaches underpin different programming paradigms and have their own strengths and weaknesses.

The SIMD architecture

SIMD-based computers employ a large amount of interconnected *processing elements* (PE) running the same programming instructions concurrently but working with different data (Figure 7.13). Each PE has its own local processing memory, but the nature of the PEs and the mode of communication between them varies for different implementations. PEs are usually simpler than conventional processing units used on SISD machines because PEs do not usually require individual instruction fetch mechanisms; this is managed by a master control unit (MCU). Communication between PEs commonly involves an orthogonal grid of data pathways and each PE can often communicate directly with its four or eight nearest neighbours.

SIMD computers are essentially synchronous models because all PEs execute the same instruction at the same time. The only control the programmer has over an individual PE is to allow or prevent it from executing the instruction. This makes SIMD computers easier to program than MIMD computers (see below) because the control of different programs

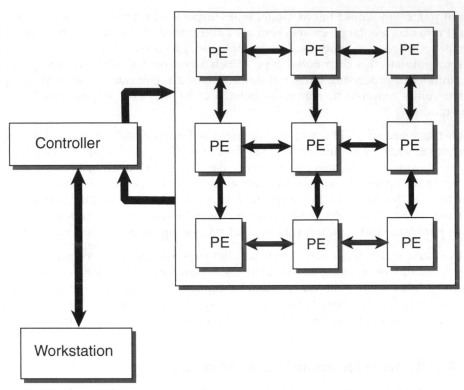

Figure 7.13 The SIMD architecture employs a large amount of interconnected processing elements (PE) running the same programming instructions concurrently but working on different data

running on each processor of a MIMD machine becomes very demanding as the number of processors increases. SIMD architectures therefore suit the provision of very large arrays of PEs; current SIMD computers may have as many as 65 536 PEs.

The MIMD architecture

MIMD-based computers employ independent processors running different programming instructions concurrently and working on different data. There are a number of different ways to construct MIMD machines, according to the relationship of processors to memory and to the topology of the processors.

Regarding the relationship between processor and memory, we identify two fundamental types: *shared global memory* and *private local memory*. In shared global memory, each processor has a direct link to a single global memory via a common pathway, or *bus* in computer science jargon (Figure 7.14). In this case, processors only communicate with each other via the global memory. This type of MIMD architecture is sometimes preferable because they are relatively straightforward to program. The major drawback is that the number of processors should be limited to accommodate the capacity of the pathway.

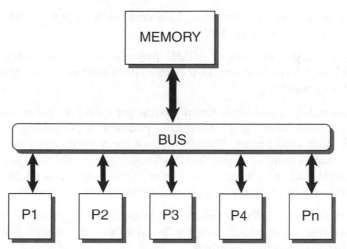

Figure 7.14 The processors of MIMD-based computers can only communicate with each other via the global memory

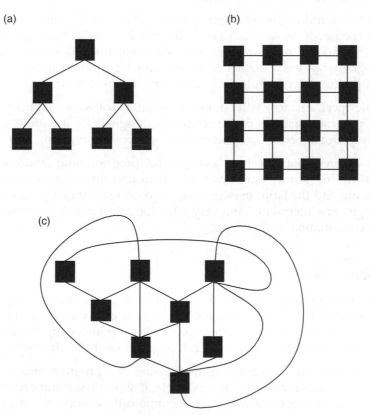

Figure 7.15 The most often used topologies of MIMD transputers are binary trees (a), two-dimensional grids (b) and hypercubes (c)

Otherwise, too many processors competing to access the memory would need complicated mechanisms to prevent data traffic jams.

In order to employ large amounts of MIMD processors it is necessary to provide each processor with its own local memory. In this case, each unit is normally encapsulated in a microchip called a *transputer*.

Since MIMD transputers execute different instructions and work on different data, the interconnection between them must be carefully planned. It is unthinkable to connect each transputer to all other transputers. The number of connections rises as the square of the number of units. From a number of possibilities, most customarily used topologies are *binary trees*, *two-dimensional grids* and *hypercubes* (Figure 7.15). Sophisticated MIMD implementations allow for user-configuration of topologies to suit particular needs.

MIMD computers are essentially asynchronous because each processor or transputer may proceed at its own rate, independently of the behaviour of the others. They may, of course, perform in synchronous mode if required.

7.2.2 Parallel programming strategies

It seems likely that within a few years computers will be based upon a parallel architecture of some kind. Some of these machines will probably make their parallel architecture transparent to high-level programmers and will somehow allow for the processing of sequential programs as well. In this case, sequential programs must be adapted either manually by a programmer or automatically by the machine itself.

Best achievable performance will certainly require software explicitly designed for parallelism. A program that has been specifically designed for a parallel architecture will unquestionably perform better than a converted sequential program.

There are two different but closely related parallel programming strategies: *decomposition* and *parallelisation*. The former employs methods to decompose a sequential program for parallel processing and the latter uses techniques to design parallel programs from scratch. All these strategies are interrelated and very often the solution to a problem is best achieved by applying a combination of them.

Decomposition

The main objective of decomposition is to reduce the execution time of a sequential program. In general, it is possible to split a sequential program into various parts, some of which would have good potential for concurrent processing. By allocating these parts to various parallel processors, the runtime of the overall program could be drastically reduced.

Decomposition requires caution because the splitting of a program into a large number of processors incurs considerable costs. For example, it should be estimated that individually processed parts might require demanding communication mechanisms to exchange data between them. That is, a large number of processors does not necessarily lead to good performance.

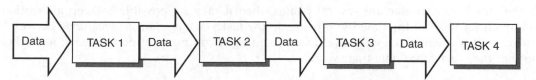

Figure 7.16 In pipeline decomposition, each input data passes through each of the modules sequentially but as they move through the pipeline the modules work simultaneously

We identify three main strategies for decomposition: *trivial*, *pipeline* and *geometric*. A trivial decomposition takes place when the original sequential program processes a large amount of data with no interdependencies. In this case, there is no major technical problem to split the task, since parallelism may be achieved simply by running copies of the entire program on various processors concurrently. Conversely, a pipeline decomposition strategy is applied when a program can be split into various modules and allocated to different processors. All input data still passes through each of the modules sequentially but as they move through the pipeline the modules work simultaneously (Figure 7.16). The first input element passes through the first stage of the pipeline and, after being processed, it then passes on to the next stage of the pipeline. Whilst this first element is being processed at the second stage, the second input element is fed into the first stage for processing, and so on. All stages work simultaneously because each has its own exclusive processor. Finally, geometric decomposition splits the data rather than the program. This strategy is suited for those problems where identical operations are applied to different parts of a large data set. As with the case of trivial decomposition, each processor has a copy of the whole program. The difference between trivial and geometric decomposition is that the latter is designed to cope with data dependencies; that is, the result of the operation of one data subset may require information about the operation of other data subsets. Geometric decomposition works best when the data subsets need to interact with their neighbouring subsets. In this case the data is distributed to a certain number of processors arranged in a lattice and the program will have to be provided with mechanisms for passing messages across the processors.

Parallelisation

There are two main different approaches to the design of a parallel program: *specialist processing* and *task farm processing*. These approaches have many links to the decomposition categories introduced above and both concepts certainly complement each other.

Specialist processing

Specialist processing achieves parallelism by dividing the problem into specific types of tasks. For example, if the problem is to build a piano, then there will be a variety of specialised tasks such as making the keys, strings, case, etc. Once all pieces are complete, the piano is then assembled. Specialist workers are assigned for each task in order to ensure the quality of the components and increase the speed of construction. However, if the objective is to build only one piano, workers may not be able to start working simultaneously because some will have to wait for other colleagues to finish in order to start their job. Conversely, if the aim is to manufacture several pianos, then it may be possible to keep all workers occupied for most of the time by 'pipelining' the tasks (see *pipeline decomposition* above). In any case, it is essential to establish efficient channels of communication between the workers so that they can coordinate their work.

Task farm processing

In task farm processing, the problem is divided into several tasks but not necessarily targeted for specialist processors. To continue the above-mentioned metaphor of the piano, there are no specialist workers in the team here because it is assumed that all of them are capable of carrying out any of the tasks for building a piano. In this case, there is a foreman who holds a list of tasks to be performed. Each worker takes a task and when it is accomplished he or she comes back and selects another. Communication occurs mostly between a worker and the master; for the sake of maximum performance, workers are not encouraged to talk to each other because their attention must not be diverted from the main task.

7.2.3 Superconductivity, holographic memory and optical pipelines

Besides the parallelisation techniques discussed above, technology is also progressing towards improved processing units, memory devices and connectivity. RSFQ (Rapid Single Flux Quantum) technology based on superconductivity seems to be a promising alternative to current CMOS (Complementary Metal Oxide Semiconductor) processors. Super-conductivity is the ability of some materials to conduct electricity with no resistance when cooled to very low temperatures: a loop of a superconducting wire could, in principle, sustain a running electric current forever. RSFQ processors use an extremely fast swtiching mechanism based on superconducting loops. These loops, called SQUIDs (short for Superconducting Quantum Interference Devices) employ two non-linear switching devices called *Josephson junctions*. These loops can operate in distinctive ways: they may contain no current, they may sustain the basic current, or they may have a current that is some multiple times the basic current, but with nothing in-between; that is, current values are discrete. RSFQ takes advantage of this property to represent the 0s and 1s of digital codification as discrete current rather than as distinct voltage levels. RSFQ gates can operate at processing speeds of approximatety 800 gigahertz, that is about 100 times quicker than CMOS gates.

Holographic memory is another innovation that will probably replace semiconductor-based DRAM (Dynamic Random Access Memory). Holographic memory uses light-sensitive materials to accumulate large blocks of data. Storage capacities of 10 gigabits in blocks as small as a few cubic centimeters can be achieved by means of *photorefractive storage*. In photorefractive storage, a plane of data modulates a laser beam signal with a reference beam in a tiny block of lithium niobate (a storage material). The hologram results from the effect that occurs when local electric fields are created by trapped, spatially distributed charge carriers excited by the interfering beams.

To connect superconducting processors and high-density holographic memory devices, researchers are developing high-capacity optical pipelines. These optical pipelines will replace electrons in metal wires with photons in fibre-optic wires employing multiple wavelengths of light (or colours) to carry digital information. This can allow for great bandwidth capacity, as separate digital signals, each with their own dedicated light wavelength, can travel together through the same channel.

7.2.4 Benefits for sound synthesis

In several ways, parallel supercomputing concepts are making their way into the synthesis technology, from the design of dedicated VLSI (Very Large Scale Integration) chips to the implementation of new paradigms for sound synthesis.

Signal processing level

Dedicated digital signal processors (called as DSP) with multiple functional units and pipeline methods have been encapsulated into VLSI chips and today are commonly found in a variety of hardware systems for sound synthesis. The functioning of a DSP is driven by a set of instructions that is loaded into its own memory from a host computer. The DSP then cycles continuously through these instructions until it receives a message to halt; a sample is produced at each cycle of the DSP. Special DSP arrangements may form arrays of DSP for simultaneously processing blocks of samples. In this case, the output for each cycle is an array of samples. Also, parallel configurations of specially designed general purpose microprocessors based on RISC (Reduced Instruction Set Computer) technology have been used on several occasions.

Software synthesis programming level

In software synthesis programming there have been a few attempts to decompose existing synthesis programming languages for concurrent processing. Pioneering research on this front has been reported by Peter Manning and his collaborators at Durham University, who in the early 1990s managed to run Csound on a parallel machine (Bailey *et al.*, 1990). Like pcmusic and Nyquist, Csound is a synthesis programming language in which instruments are designed by connecting many different synthesis units. One or more instruments are saved in a file called the *orchestra*. Then the parameters to control the instruments of the orchestra are specified on an associated file called the *score*. This score file is organised in such a way that each line corresponds to a stream of parameters to produce a sound event (or note) on one of the instruments of the orchestra. For instance:

```
i1 0 1.0 440 90
i2 1 0.5 220 80
```

In the example above, each line specifies five parameters for two different instruments, including the 'names' of the instruments (i.e. *i*1 and *i*2, respectively). Each line produces a note.

The parallel implementation of the Durham team is inspired by the task farm paradigm discussed earlier. An copy of the entire orchestra is distributed to each processor and the score is considered as a list of tasks for the processors; that is, each processor selects a line from the score for processing.

Parallel supercomputing technology is a plausible solution to computationally demanding tasks, such as physical modelling synthesis.

Synthesis modelling level

At this level a parallel computing paradigm is normally embedded in the synthesis model itself. As an example we cite an earlier version of Chaosynth (refer to Chapters 5 and 8), developed by the author in collaboration with the engineers of the Edinburgh Parallel Computing Centre, in Scotland (Miranda, 1995a).

Remember that the states of Chaosynth's CA cells represent frequency values rather than colours. The oscillators are associated to a group of cells, and in this case the frequency for each oscillator is established by the arithmetic mean of the frequencies of the cells associated to the respective oscillator. At each cycle of the CA, Chaosynth produces one grain and as the CA evolves, the components of each grain change therein. The size of the grid, the number of oscillators and other parameters for the CA are all configurable by the user. CA are intrinsically suited for parallel processing. They are arrangements of cells, each of which is updated every cycle, according to a specific global rule that takes into account the state of the

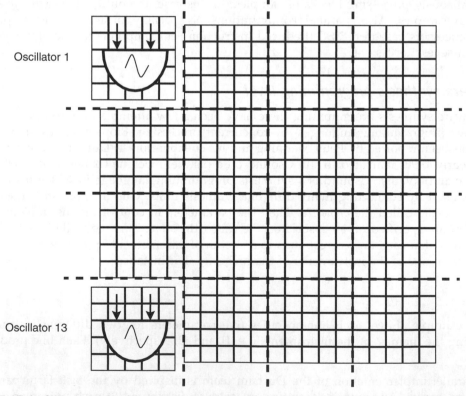

Figure 7.17 In Chaosynth, the CA grid is subdivided into a number of contiguous rectangular portions in such a way that all of them are handled concurrently by different processors. In this case, the oscillators are computed in parallel

neighbouring cells. This is a typical problem for geometrical decomposition with message passing. In this case, the grid is subdivided into a number of contiguous rectangular parts in such a way that all of them are handled concurrently by different processors. The oscillators of Chaosynth are therefore computed in parallel (Figure 7.17).

7.2.5 Do-It-Yourself supercomputer: the Beowulf architecture

Parallel supercomputers are prohibitively expensive. Most research institutions and universities cannot afford such a tremendous investment. Even when such machines are available to academic researchers, access to them often amounts to a tiny fraction of the resources shared among many other users, including heavy-duty ones who take up most of the resources for controlling things such as a nuclear power station or a military surveillance system. A music research project can hardly compete for resources in such a set-up where the priorities are heavily biased towards the needs of the major sponsors. One way to get round this problem is to build a parallel computer by linking inexpensive personal computers.

Back in 1994, Thomas Sterling and Don Becker created a cluster computer using 16 standard personal computers each containing an Intel 486 processor connected via Ethernet at CESDIS (Centre of Excellence in Space Data and Information Sciences) in the USA. The Linux operational system was used to run the cluster. Linux, a Unix-like operational system originated by Linus Torvalds in the early 1990s, is ideal for managing such clusters (Sobel, 1997). Sterling and Becker called their machine *Beowulf* and their idea of using off-the-shelf processors to build parallel computers soon spread through academic and research communities worldwide. Nowadays, Beowulf is a label for a technology for building massive parallel systems using a network of single PCs. Many universities and research centres around the world are recycling their obsolete PCs to build Beowulf systems. For example, the Oak Ridge National Laboratory in the USA has recently reported that they have built a Beowulf cluster of 130 PCs. One of them serves as the front-end node for the cluster. It features two Ethernet cards, one for communicating with users and the other for talking with the rest of the units. The system divides the computational workload into many tasks, which are assigned to the individual units. The PCs coordinate their tasks and share intermediate results by sending messages to each other.

Anyone can construct a parallel computer using PCs. Most research laboratories use PCs with dual booting schemes so that they can be rebooted into either Linux or Microsoft Windows. One could set them up so that when they boot into Linux they are nodes of a Beowulf cluster and when they boot into Windows they are just standard desktop machines. The minimum requirements to serve as a node of a Beowulf cluster are that the PC should contain at least an Intel 486 CPU and motherboard, at least 16 MB of memory, a minimum of 200 MB of disk space, an Ethernet interface, a video card and a floppy disk drive. All this gear should be supported by Linux. Keyboards and monitors are required only for the initial loading of the operational system and configuration, unless the individual machines are also being used for other purposes. In additon to Linux, some additional pieces of software are needed in order to run parallel programs, such as MPI (Message Passing Interface) software. This software is usually available at no cost from the Internet. The good news here is that parallel codes developed on Beowulf systems using standard MPI libraries can be run on commercial parellel supercomputers without major modifications.

Although a Beowulf cluster may not outperform the fastest parallel supercomputers in the world, the idea that you can build relatively inexpensive parallel computer environments that can be fully dedicated to your own research agenda is surely highly attractive.

More information on how to build a Beowulf computer can be found in the book *How to Build a Beowulf: A Guide to the Implementation and Application of PC Clusters*, featuring chapters by Sterling, Becker and many others (Sterling *et al.*, 1999).

7.3 Evolutionary sound synthesis

A recurring theme in this book is the difficulty to set a synthesis algorithm with the right parameter values for producing a desired sound. The first section of this chapter described an approach for designing interfaces that can translate intuitive sound descriptors into the numerical values of a given synthesiser. This approach is largely based upon an area of AI research known as knowledge-based systems. By and large, the idea is to furnish the synthesiser with an interface that facilitates the exploration of a vast range of possibilities. In this case, the intelligence of the system comes from its embedded knowledge about synthesis, which drives the exploration in specific ways. This is generally known in AI research as a heuristic search.

Back in 1993 Andrew Horner and his colleagues proposed a different approach to searching for synthesis parameter values (Horner *et al.*, 1993). They employed an optimisation technique called *genetic algorithms* to find the parameters for an FM synthesiser (Chapter 2) that could produce a sound whose spectrum matched the spectrum of a given sound example as closely as possible. Obviously, the objective of this particular system was not to guide the user in creative exploration of the parametrical space of FM synthesis, but it provided good insight into the potential of a relatively new computational paradigm in the realm of sound synthesis: *evolutionary computation*.

On the whole, evolutionary computation brings together a pool of algorithms for searching for one or more solutions to a problem in a vast space filled with possible solutions. In this context, the problem of finding the right synthesis parameters for producing a sound could be defined as a search for a point in a multi-dimensional space, whose co-ordinates correspond to the respective parameters of the synthesiser in question.

Evolutionary search algorithms are much inspired by Charles Darwin's evolutionary theory introduced his famous book *The Origins of Species by Means of Natural Selection*, first published in 1859 (Miller and van Loon, 1990), hence the term 'evolutionary computation'. Indeed, these algorithms seem to search by evolving solutions to problems, because instead of testing one potential solution at a time in the search space, these algorithms test a population of candidate solutions. This evolutionary metaphor is extremely interesting because evolution as such is not explicitly programmed. Rather, it is an emergent property of the algorithm. The computer is instructed to maintain populations of solutions, allow better solutions to reproduce, and have bad solutions die. The offspring solutions inherit their parents' characteristics with some small random variation, and then the best solutions are allowed to reproduce, while the bad ones die, and so forth. After a number of generations, the computer will have evolved solutions which are substantially better than their long-dead ancestors.

Evolutionary algorithms require guidance to direct the evolution towards better areas of the search space. They receive this guidance by evaluating every solution in the population in order to determine its *fitness*. The fitness of a solution is a score based on how well the solution fulfils the problem in question. This score is calculated by means of a *fitness function*.

The following paragraphs present a brief introduction to two evolutionary computing paradigms that are beginning to make their way into sound design systems: *genetic algorithms* and *genetic programming*.

7.3.1 Genetic algorithms

The genetic algorithm, or GA, is perhaps the most popular evolutionary technique amongst musicans. The basic idea behind GAs first appeared in the early 1960s, but John Holland is often quoted as one of the pioneers of GAs for his work on modelling adaptive natural systems (Holland, 1975).

GAs use two separate spaces: the search space and the solution space. The former is a space of coded solutions of the problem, or *genotypes* in GA jargon, and the solution space is the space of actual solutions, or *phenotypes*. Genotypes must be mapped onto phenotypes before the quality (or fitness) of each solution can be evaluated (Figure 7.18). GAs mantain a population of individuals where each individual consists of a genotype and a corresponding phenotype. Whilst phenotypes usually consist of a collection of parameters, genotypes consist of coded versions of these parameters. Coded parameters are normally called *genes*. The collection of genes of a *genotype* is often stored in a memory as a string, and it is referred to as the *chromosome*.

GA theory uses jargon and a number of metaphors drawn from the biological sciences but readers should not be put off: the principles behind the jargon and methaphors are fairly straightforward.

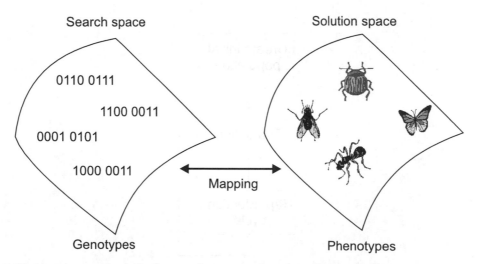

Figure 7.18 Mapping genotypes in the search space to phenotypes in the solution space

GAs are normally used for finding optimal solutions to engineering and design problems where many alternative solutions may exist, but whose evaluation cannot be ascertained until they have been implemented and tested. In such cases, the optimal solution can be sought by manually generating, implementing and testing the alternatives or by making gradual improvements to alleged non-optimal solutions. As the complexity of the problem in question grows, however, these procedures become increasingly unsatisfactory and wasteful. Hence the need for an automated method that explores the capability of the computer to perform many cumbersome combinatorial tasks. Notwithstanding, genetic algorithms go beyond standard combinatorial processing as they embody powerful and yet simple mechanisms for targeting only potentially fruitful combinations.

The sequence of actions illustrated in Figure 7.19 portrays a typical genetic algorithm. At the beginning, a population of abstract entities is randomly created. Depending on the application, these entities can represent practically anything, from the fundamental components of an organism, to the commands for a robot, or the parameters of a sound synthesiser. Next, an evaluation procedure is applied to the population in order to test whether it solves the task or problem in question. As this initial population is bound to fail the evaluation at this stage, the system embarks on the creation of a new generation of entities. Firstly, a number of entities are set apart from the population according to some prescribed criteria. These criteria are often referred to as *fitness for reproduction* because this sub-set will undergo a mating process in order to produce offspring. The fitness criteria obviously vary from application to application but in general they indicate which entities from the current generation work best. The chosen entities are then combined (usually in pairs) and give birth to the offspring. During this reproduction process, the formation of the offspring involves a mutation process. Then, the offspring are inserted in the population, replacing their parents. The fate of the remaining entities of the population not selected for reproduction may vary, but they usually die out without causing any effect. At this point we say that a new generation of the population has evolved. The evaluation procedure is now applied to the new generation. If the population still does not meet the

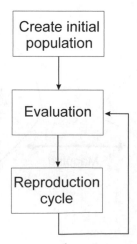

Figure 7.19 A typical genetic algorithm arrangement

objectives, then the system embarks once more on the creation of a new generation. This cycle is repeated until the population passes the evaluation test.

In practice, GAs normally operate on a set of binary codes that represent the entities of the population. These operations involve three basic classes of processes: *recombination, mutation* and *selection*. Whilst the recombination process leads to the exchange of information between a pair of codes, the mutation process alters the value of single bits in a code. Recombination produces offspring codes by combining the information contained in the codes of their parents. Depending on the form of the representation of the codes, two types of recombination can be applied: *real-valued recombination* or *binary-valued crossover*.

In order to illustrate a typical genetic algorithm in action, consider a population P of n entities represented as 8-bit codes as follows: $P = \{11010110, 10010111, 01001001, \ldots\}$. Then, suppose that at a certain point in the evolutionary process, the following pair of codes is selected to reproduce: p_7: 11000101 and p_{11} = 01111001. Reproduction in this example is defined as a process whereby the couple exchanges the last three digits of their codes followed by a mutation process. In this case, the pair p_7 and p_{11} will exchange the last three digits of their codes as follows:

p_7:   11000[101] → 11000[001]
p_{11}: 01111[001] → 01111[101]

Next, the mutation process takes place; mutation usually occurs according to a probabilistic scheme. In this example, a designated probability determines the likelihood of shifting the state of a bit from zero to one, or vice versa, as the code string is scanned. Mutation is important for introducing diversity in the population, but one should always bear in mind that higher mutation probabilities reduce the effectiveness of the selective process because they tend to produce offspring with little resemblance to their parents. In this example, the third bit of p_7 and the fourth bit of p_{11} were mutated:

p_7:   11[0]00001 → 11[1]00001
p_{11}: 011[1]1101 → 011[0]1101

The new offspring of p_7 and p_{11} are 11100001 and 01101101, respectively.

Codification methods

In order to make effective use of genetic algorithms, one has to devise suitable methods both to codify the population and to associate the behaviour of the evolutionary process with the application domain, which in our case is sound synthesis. The rule of thumb is that one should try to employ the smallest possible coding alphabet to represent the population.

A typical codification method is the *binary string* coding whereby each individual is represented by a string of some specified length; the 8-bit coding illustrated above is a typical example of the binary string coding. This coding method is interesting because each digit of the code, or groups of digits, can be associated with a different attribute of the individual (Figure 7.20).

A number of variations of the binary string coding may be devised. For example, one could devise codes using large binary strings divided into words of a fixed length; each word

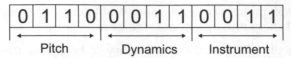

Figure 7.20 Each digit of the binary string can be associated to a different attribute of the individual being coded; e.g. the attributes of some sound being synthesised

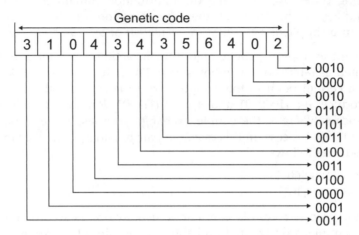

Figure 7.21 Each digit of the decimal string is computed at the binary representation level

corresponding to a decimal value. In this case, the code is a decimal string but the actual computing of the genetic algorithm processes occurs at their corresponding binary representation (Figure 7.21).

Selection mechanisms

The mechanism for selecting the population subset for reproduction varies according to the application of the genetic algorithm. This mechanism generally involves the combination of a fitness assignment method and a probability scheme. One of the simplest selection mechanisms is the *stochastic sampling selection*. In order to visualise how this selection mechanism works, imagine that the whole population is projected onto a line made of continuous segments of variable lengths. Each individual is tied to a different segment whose length represents its fitness value. Then, random numbers whose maximum value is equal to the length of the whole line are generated; the individuals whose segments cover the values of the random numbers are selected. A variation of the stochastic sampling selection would be to superimpose some sort of metric template over the line. The number of measurements on the template corresponds to the amount of individuals that will be selected and the distance between the measures is defined as $d = 1/n$, where n is the number of measures. The position of the first measure is randomly decided in the range of $1/n$.

Another widely used selection mechanism is the *local neighbourhood selection*. In this case, individuals are constrained to interact only within the scope of a limited neighbourhood. The neighbourhood therefore defines groups of potential parents. In order to render this process more effective, first the algorithm selects a group of fit candidates (using stochastic sampling, for example) and then a local neighbourhood is defined for every candidate. The mating partner is selected from within this neighbourhood according to its fitness criteria. The neighbourhood schemes that are commonly used are ring, two-dimensional and three-dimensional neighbourhoods, but more complex schemes may also be devised. The distance between neighbours does not necessarily need to be equal to one and not all immediate neighbours need to be considered (Figure 7.22).

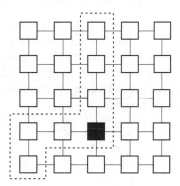

Figure 7.22 Partial two-dimensional neighbourhood with distance = 3

Many more selection methods exist but it is impracticable to survey all of them here. Successful applications of genetic algorithms often use selection mechanisms specially tailored to the task at hand.

Typically, genetic algorithms are used for optimisation purposes. Thus the selection operation is responsible for choosing the best possible codes according to a certain predetermined *fitness criterion*. There are applications, however, in which the user is required to interact with the system during the search process. In the case of sound design software, for example, a number of synthesised sound candidates may be presented to the user, who then assigns fitness rankings to them based on how close these sounds are to the sound target.

7.3.2 Genetic programming

In essence genetic programming, or GP, embodies a specific way of using genetic algorithms for handling special types of problems such as evolvable software and hardware. Because these problems require slightly more sophisticated mechanisms and operators, scientists tend to consider GP as a whole research domain in its own right.

The main mentor of GP is John Koza who originally intended to use evolutionary computing to actually evolve computer programs automatically (1992).

Similarly to GA, GP starts with a population of randomly generated computer programs. The set of functions that may appear at the internal points of a program tree may include ordinary arithmetic functions and conditional operators. The set of terminals appearing at the external points typically includes the program's external inputs (e.g. *X* and *Y*) and random constants (e.g. 3.2 and 0.4). The randomly created programs typically have different sizes and shapes.

At each cycle of the GP run, each individual program in the population is evaluated to determine how good (or 'fit') it is at solving the problem at hand. Programs are then selected from the population based on their fitness to participate in the various genetic operations, as introduced in the previous section. After many generations, a program may emerge that solves, or approximately solves, the problem at hand.

The main difference between GAs and GP is that the latter does not make a distinction between the search space (genotypes) and the solution space (phenotypes), in the sense that GP manipulates the solutions themselves rather than their coded versions. Also, solutions are represented in GP in a layered fashion (Figure 7.23). This layered, or tree-like, representation facilitates the application of basic evolutionary operations, such as crossover and mutation, to strings of various lengths.

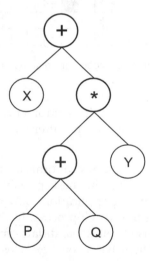

Figure 7.23 Genetic programming was originally developed to evolve computer programs that can be represented hierarchically

The GA crossover operator works in a slightly different manner here. For example, consider two strings as follows: 'A + B/C' and '– A/B + C'. If we use the standard GA crossover operation at point 1, then the two children are: '– + B/C' and 'A A/B + C'. Clearly, this operation produces invalid expressions. If the strings were arranged in a layered tree structure, the crossover could be applied to interchange the branches of the trees without disrupting the syntax of the expressions (Figure 7.24).

GP is suitable for evolving symbolic expressions in functional programming languages such as Lisp (a brief introduction to Lisp is available in Chapter 8). For example, the two parent solutions used to illustrate the crossover operator in Figure 7.24 could be written in Lisp as

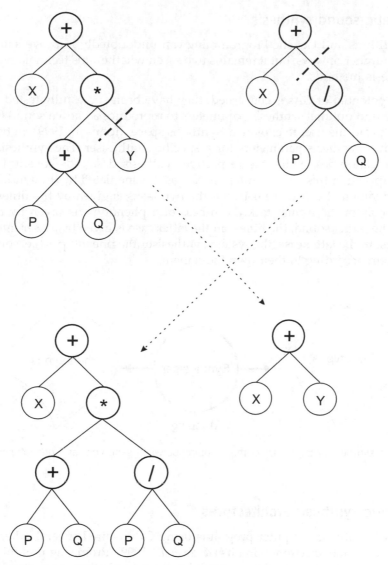

Figure 7.24 The behaviour of the crossover operator in GP

follows: (+ A (/ B C)) and (+ (/(– A) B) C). As a matter of fact, Nyquist (available on the accompanying CD-ROM) is ideal for implementing GP-evolved instruments.

In addition to the classic GA crossover and mutation operations, GP includes *architecture-altering* operations such as *subroutine duplication, argument duplication, subroutine creation, subroutine deletion* and *argument deletion*. A presentation of these operations is beyond the scope of this book. Readers interested in studying them are invited to refer to the appropriate literature; a good introduction can can be found in John Koza's book *Genetic Programming: On the Programming of Computers by Means of Natural Selection* (1992).

7.3.3 Genetic sound synthesis

Genetic Algorithms and Genetic Programming can undoubtedly improve sound synthesis systems in a number of ways, but it remains to be seen whether this technology can also lead to new synthesis methods.

As far as the potential of GAs is concerned, they have been succesfully tested in two kinds of tasks: (a) to find optimal synthesis parameters to reproduce a known sound (Horner *et al.*, 1993) and (b) to aid the user to explore a 'synthesis space' (Johnson, 1999). In both cases, the GA manipulates synthesis parameters for a specific synthesiser. The synthesis parameters constitute the genotypes, whilst the respective synthesised sounds constitute the phenotypes. The mapping in this case is rather special, as it is mediated by the synthesiser (Figure 7.25). The fundamental difference between the two tasks is given by the fitness evaluation process. Whereas in the former task the fitness of a phenotype is given by the degree of similarity to the target sound, the fitness in the latter case is given by a user interacting with the algorithms. In the latter case, the system synthesises the phenotypes (i.e. sounds) and the user ranks them according to their own judgement.

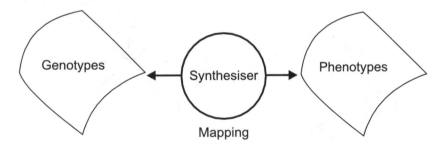

Figure 7.25 The synthesiser takes care of the mapping between genotypes and phenotypes

7.3.4 Genetic synthesis architectures

The concept of evolving computer programs using GP inspired several other applications such as the evolution of electronic circuits (Koza *et al.*, 1999). Promising preliminary work on evolving synthesiers was recently reported by Ricardo Garcia at the 1st Workshop on Artificial Life Models for Musical Applications (ALMMA) held in Prague in September 2001 (Bilotta *et al.*, 2001).

As we have seen in the previous chapters, the architecture of a synthesiser is normally represented as a network of interconnected unit generators (Figure 6.9, Chapter 6). We could say that the synthesis space is the space spanned by all the possible combinations of unit generators and their connections. This space should, however, be restricted to a countable, and preferably small, number of different units. Also a certain limit should be imposed as to the total amount of units that an architecture can hold. And indeed, a valid architecture should not contain any dangling connections; i.e. all units must be fully and appropriately connected.

Garcia limited the number of units in his experimental system to nine, consisting of four types: *Type A* (units with two inputs and one output), *Type B* (units with one input and two outputs), *Source* (1 output) and *Render* (1 input):

- Add (Type A)
- Mult (Type A)
- Filter + Oscillator (Type A, oscillator with filtered output)
- Oscillator (Type B, can produce sine, triangular and square waves)
- Filter (Type B, two resonance peaks)
- Split (Type B, splits a signal into two streams)
- Constant (Type B)
- Source (Source)
- Render (Render)

Bear in mind that these units constitute the entire repertoire of the synthesis programming language at hand here and some of them are not unit generators in the sense considered in Chapter 1; e.g. *Add* and *Mult* are operations.

Searches in the space of synthesis architectures are carried out by means of three operators: *addition, deletion* and *displacement of blocks and their connections*. The problem with the standard representation of synthesisers, as illustrated in Figure 6.9, is that it does not facilitate the application of the GP operations. It is very complicated to try to keep the architecture valid even after very simple operations. For example, if a new oscillator block is to be added, there is no guarantee that this new block will be placed coherently.

A similar problem was encoutered by Koza when he tried to use GP to evolve electronic circuits. Koza's solution to this has become standard in most GP applications and Garcia's is no exception: the solution is to devise a tree-like representation for the circuits. This idea comes from discrete mathematics, more precisely from graph theory. In a very wide sense, graphs can be thought of as visual representations of the mathematical relations between objects. A graph is normally represented as a collection of points, called vertices, some of which are connected by lines, called edges (Vince and Morris 1990). What interests us here is the fact that a synthesiser's architecture can be represented as a graph and certain kinds of graphs can be represented as trees, another visual mathematical structure (Figure 7.26).

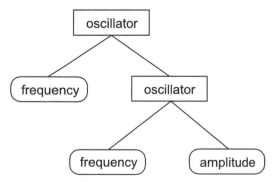

Figure 7.26 A tree representaion of a synthesiser

With the tree representation of synthesis architectures using the set of units listed above, one can apply the operations to the branches of the tree. This is a big improvement to the problem of maintaining the coherence of the evolving synthesiser's architecture.

As for using the fitness function to evaluate the performance of the synthesisers, Garcia employed a distance measure that compares the magnitude spectrogram of two sounds (Garcia, 2001). The synthesiser that produces the sound that best matches the spectrum of a target spectrum is considered to be the fittest.

8 Introduction to the software on the accompanying CD-ROM

On the accompanying CD-ROM there are a number of software synthesis systems in various styles, ranging from generic programming languages to programs dedicated to specific synthesis techniques. This chapter gives an overview of these systems, but the reader is strongly advised to refer to the documentation on the CD-ROM for more in-depth information on each of them. The documentation for most of the programs includes comprehensive tutorials in addition to the user guide.

A number of the systems on the CD-ROM are fully working versions, but those of a commercial nature may have been abridged according to the policy of each manufacturer. As far as this book is concerned, any should be able to run at least the examples cited in this book.

8.1 pcmusic

The synthesis programming language pcmusic is the DOS version of cmusic for PC-compatible computers which was originally designed to run on Unix machines. The cmusic language is a classic Music N type of synthesis programming language devised in 1980 by F. Richard Moore at the University of California, San Diego, USA (Moore, 1990).

The system is provided with a large library of unit generators and signal processing subroutines that have evolved over the years. It works as a compiler, therefore it was not designed to work in real time. There are virtually no limitations to the complexity of the instruments in pcmusic because the system does not bother with processing speed. Slower processors naturally take longer to compile, but faster ones do not necessarily improve the instrument design capabilities of the language.

In classic Music N types of programming languages such as pcmusic, the electronic musician designs instruments by writing instructions specifying interconnections of unit generators and setting up the lookup tables containing the waveforms for oscillators, envelopes and other controllers. Once the instruments are specified, then the musician must put together a list of notes to play them. Each note of the list consists of start time and duration specifications, followed by a number of synthesis parameter values, as required by the instrument that will play it. The term 'note' here refers to a generalised sound event of any complexity. Both instruments and notes are normally saved onto one single file, called *the score*. The values for global synthesis parameters, such as sampling rate and number of channels, must be specified in a file called *Pcmusic.ini* normally found in the current working directory. This file contains pcmusic-specific information about the configuration of the system. Novices are therefore advised not to make major changes to the *Pcmusic.init* file unless they are completely sure of what they are doing.

A pcmusic score is typically divided into three major sections:

- Instrument specifications
- Lookup table definitions
- Note list

8.1.1 The specification of instruments

In this section one defines the architecture of the instruments by specifying a certain configuration of unit generators. The general form of an instrument is as follows:

```
instrument <start time> <name>;
<unit generator instruction 1>;
<unit generator instruction 2>;
. . .
<unit generator instruction n>;
end;
```

Every instrument begins with the instruction *instrument* followed by a starting time (normally set to zero) and a label, and closes with the instruction *end*. The body of the instrument contains unit generator instructions. Each instruction line begins with the name of the respective unit generator (e.g. *osc* for an oscillator; *out* for audio output; *flt* for a filter, etc.) and it is followed by a list of parameters. The following example shows the codification of a simple instrument:

```
instrument 0 simple;
osc b1 p5 p6 f1 d;
out b1;
end;
```

The instrument, labelled as *simple*, has an oscillator (*osc*) which outputs samples to the audio output unit (*out*) via a variable called *b1*. The amplitude and frequency values for the oscillator will be specified in the note list via two variables, *p5*, and *p6*, respectively. The waveform for this oscillator is referred to as *f1* and it should be specified elsewhere in the score. The last parameter of the oscillator is related to its initial phase and the 'value' *d* is usually employed. It is not necessary to concentrate on this parameter at this stage; more details about it can be found in Robert Thompson's tutorial on the accompanying CD-ROM.

The syntax of pcmusic is relatively simple. Each instruction line of a program always begins with a pcmusic command and closes with a semicolon (;). A number of parameter fields are arranged between the initial command and the final semicolon. Each parameter field may be a single value, a variable or an expression, which are separated by blanks or commas. Note that a parameter field itself cannot contain blank spaces.

8.1.2 The specification of lookup tables

The pcmusic package has a number of lookup table generators referred to as *gen* followed by a number; e.g. *gen1* (generates a straight line), *gen4* (produces exponential curves), *gen5* (generates sinusoidal partials), to cite but a few. The instruction *generate* is used to activate a certain *gen* subroutine, as follows:

```
generate <time> <gen subroutine> <table name> <list of parameters>;
```

Two examples using *gen5* to produce sinusoidal components are given as follows:

```
generate 0 gen5 f1 1,1,0;
generate 0 gen5 f2 1,1,0 5,1/2,0;
```

The generator *gen5* produces a combination of sinusoidal partials described by triplets of parameters specifying the harmonic number, the relative amplitude of the harmonic and its phase offset, respectively. The first example generates a lookup table called *f1* containing a single sinewave; i.e. harmonic number = 1, relative amplitude = 1 and phase offset = 0. The second example generates table *f2* containing a waveform resulting from the addition of two sinewaves: a fundamental and its fifth harmonic with halved amplitude.

8.1.3 The specification of note lists

The format of the specification of note lists is as follows:

```
note <start time> <instrument> <duration> <synthesis parameters>;
note <start time> <instrument> <duration> <synthesis parameters>;
note <start time> <instrument> <duration> <synthesis parameters>;
etc. . ..
```

The following example illustrates a sequence of four notes for the instrument *simple* above. Note the use of explicit symbols for units such as dB and Hz for amplitude and frequency values, respectively. Start time and duration values are specified in seconds.

```
note 0 simple 1 -6 dB 370.0 Hz;
note 1 simple 1 -6 dB 425.3 Hz;
note 2 simple 2 -3 dB 466.2 Hz;
note 4 simple 2 0 dB 622.3 Hz;
```

8.1.4 A complete example

A complete pcmusic score can now be put together using the examples given above. The *terminate* instruction is used to indicate the end of the score. Pcmusic allows for insertion of comments within brackets in the code:

```
{ ------------------------------------
This is an example of a simple
instrument in pcmusic.
------------------------------------
Initialisation:
------------------------------------}
#include <cmusic.h>
#include <waves.h>
{ ------------------------------------
Instrument specification:
------------------------------------}
instrument 0 simple;
osc b1 p5 p6 f1 d;
out b1;
end;
{ ------------------------------------
Lookup tables: f1 is a sinewave
------------------------------------}
generate 0 gen5 f1 1,1,0;
{ ------------------------------------
Note list:
------------------------------------}
note 0 simple 1 -6 dB 370.0 Hz;
note 1 simple 1 -6 dB 425.3 Hz;
note 2 simple 2 -3 dB 466.2 Hz;
note 4 simple 2 0 dB 622.3 Hz;
terminate;
```

One of the most interesting features of pcmusic is that it has a pre-processor built into its compiler. When the compiler is activated, the score is pre-processed before the compilation takes place. Lines beginning with *#include* instruct the pre-processor to include a copy of a file at that point of the code. In the example above, the pre-processor will include the files *Cmusic.h* and *Waves.h*. These files contain the code for a variety of subroutines that the system needs for the compilation of the instrument; e.g. the code for the unit generators.

The pre-processor also watches for *#define* instructions, which are used to create symbolic labels for synthesis parameters. For example, the statements

```
#define AMP p5

#define FREQ p6
```

instruct the compiler that the symbols 'AMP' and 'FREQ' can be used anywhere in the score in place of *p5* and *p6*, respectively:

```
osc b1 AMP FREQ f1 d;
```

Finally, in addition to being able to synthesise sounds from scratch using oscillators and mathematically created lookup tables, pcmusic also has the ability to incorporate recorded sounds for manipulation. This leads to a wide range of synthesis possibilities not otherwise available on some synthesis systems.

On the CD-ROM the reader will find a very comprehensive tutorial and a wide variety of examples specially prepared for this book by the composer Robert Scott Thompson of Georgia State University in Atlanta, Georgia, USA.

8.2 Nyquist

Nyquist, developed by Roger Dannenberg at the Carnegie Mellon University, USA, is a programming language for music composition and sound synthesis: it supports both high-level compositional tasks and low-level signal processing within a single integrated environment.

Nyquist is implemented on top of a programming language called Lisp, more specifically XLISP, an extension of Lisp that is freely available for non-commercial purposes. Nyquist can be thought of as a library of Lisp fuctions that can be called up by other user-specified functions in a program. In fact, XLISP has been extended to incorporate sounds as a built-in data type, which gives Nyquist a great deal of expressive power. Composers thus program their synthesisers in Nyquist as if they were writing Lisp programs. The great advantage of this is that composers have the full potential of Lisp, which is a powerful language used in Artificial Intelligence research, combined with a great variety of functions and abstract structures for composition devised by Dannenberg. The caveat of Nyquist is that composers should be familiar with Lisp programming in order to take full advantage of it. A brief introduction to Lisp is given below, but readers wishing to use Nyquist more effectively are encouraged to study further.

Helpful documentation for both Nyquist and XLISP are given in the Nyquist Reference Manual in folder *nyquist*. Also, good introductory articles and a short tutorial can be found in *Computer Music Journal*, 21:3, published in 1997.

8.2.1 Basics of Lisp programming

Lisp presents itself to the user as an interpreter; it works both as a programming language and as an interactive system: it waits for some input or command from the user, executes it, and waits again for further input. Lisp is an acronym for List Processor; almost everything in the Lisp world is a list. In programming terms, a list is a set of elements enclosed between parentheses, separated by spaces. The elements of a list can be numbers, symbols, other lists, and indeed, programs. Examples of lists are:

```
(the number of the beast)
(1 2 3 5 8 13)
(0.666 (21 xyz) abc)
(defun plus (a b) (+ a b))
```

Lisp is excellent at manipulating lists; there are functions that can do almost any operation you can possibly imagine with lists. Programs are lists themselves and they have the same simple syntax of lists. The following example illustrates what happens if one types in a short program for the Lisp interpreter. The sign '<cl>' is the command line prompt indicating that the interpreter is waiting for your command:

```
<cl>(+330 336)
666
```

When the interpreter computes (or *evaluates*, in Lisp jargon) a list, it always assumes that the first element of the list is the name of a function and the rest of the elements are the arguments that the function needs for processing. In the above example, the interpreter performs the addition of two numbers; the name of the function is the symbol '+' and the two arguments are the numbers 330 and 336.

If an argument for a function is also a function itself, then the argument is evaluated first, unless specified otherwise. The general rule is that the innermost function is evaluated first. In the case of various nested functions, the outermost one is normally the last to get evaluated. The Lisp interpreter treats each nested list or function independently. For example:

```
<cl>(+ 330 (* 56 (+ 5 1)))
666
```

In this example, the function '+' is evoked with number 330 as the first argument and with the result of the evaluation of (* 56 (+ 5 1)) as the second argument, which in turn evokes the function '*' with number 56 as the first argument and with the result of the evaluation of (+ 5 1) as the second arguments. The result of the innermost step is 6, which is then multiplied by 56. The result of the multiplication is 336 which is finally added to 330. The result of the whole function therefore is 666.

Lisp programs often involve a large number of nested functions and data. In order to keep the code visually tidy, programmers normally break the lines and tabulate the nesting. For example:

```
(+ 330
    (* 56
        (+ 5 1)
        )
    )
```

In order to write Lisp functions and programs, the language provides an instruction (or *macro*, in Lisp parlance), named *defun*; short for define function. The following example illustrates the definition of a new function labelled as *plus*:

```
<cl>(defun plus (a b) (+ a b))
PLUS
```

The first argument for *defun* (i.e. second element of the list) is the name of the new function (e.g. *plus*) followed by a list of arguments that the new function should receive when it is called up (e.g. *a* and *b*). The last argument for *defun* is the body of the new function specifying what it should actually do with the arguments. The new *plus* function defined above executes the mere sum of two arguments *a* and *b*, but it could have been a very complex algorithm or an entire program.

A variable is a symbol that represents the memory location for storing information. Variables may be used in Lisp to represent various types of information, such as a number, a list, a program, a single sound sample, a stream of samples or an entire synthesis instrument. The instruction normally used to allocate variables is *setf*. The first argument for *setf* is the label of the variable, and the second argument is the information to be associated to the variable. The following example creates a variable called *series* and associates the list (330 336 666) to this new variable:

```
<cl>(setf series '(330 336 666))
(330 336 666)
```

In Lisp, an inverted comma before a list indicates that its first element is merely a unit of an ordinary list of data, and not a command. From now on, the variable *series* can be used in place of the list (330 336 666). For example, assume that the function *first* outputs the first element of a given list:

```
<cl>(first series)
330
```

8.2.2 Introduction to Nyquist

Due to its Lisp background, Nyquist works interactively with the user via the Lisp interpreter's prompt. The symbol '>' means that the interpreter is waiting for your command. By typing the following line at the prompt we can readily make and play a sinewave:

```
> (play (osc 60 2.5))
```

The *osc* function activates a table-lookup oscillator which in this case contains a sinusoid. The first parameter 60 designates middle C as the pitch of the sound; it uses the same pitch number representation as used by MIDI systems. The second parameter determines the duration of the sound: two and a half seconds. The result of *osc* is then passed to the function *play* which is in charge of generating the sound samples. Under Windows, Nyquist writes the samples in a sound file and plays it back automatically; under other systems the user might have to use a playback application to play the files. In the simple example given above, note that the sound will inevitably begin and end with undesirable glitches because there is no envelope controlling its amplitude.

The following function defines a very simple instrument composed of an oscillator and an amplitude envelope:

```
> (defun my-instrument (pitch duration)
     (mult   (env  0.05  0.1  0.5  1.0  0.5  0.4)
             (osc pitch duration)))
```

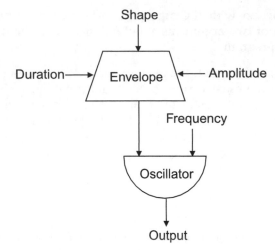

Figure 8.1 A simple synthesis instrument whereby the output of an oscillator is multiplied by an envelope in order to vary its amplitude over time

The function *env* generates a line-segment (Figure 8.1) which in turn is multiplied (function *mult*) by the output of the oscillator in order to vary the amplitude of the sound over time. We can now play this instrument by typing:

```
> (play (my-instrument c4 2.5))
```

Note that the above example uses a different notation for the pitch parameter. This is another way of saying middle C, where the number '4' indicates the octave of the note 'c'. Also, the output from *my-instrument* is much smoother than the output of previous *osc* examples due to the action of the envelope.

A simple program to make a three-note chord on *my-instrument* could be defined as follows:

```
> (defun make-chord (duration)
  (sim (scale 0.3 (my-instrument c4 duration))
  (scale 0.3 (my-instrument e4 duration))
  (scale 0.3 (my-instrument g4 duration)))))
```

The following line activates the *make-chord* program:

```
> (play (make-chord 3.5))
```

The *make-chord* program has two new pieces of information in it: *scale* and *sim*. The former is used to scale down the amplitudes of each note by 70 per cent and *sim* is a construct that tells Nyquist to produce the three events simultaneously. In a similar way to *sim*, Nyquist provides another important construct called *seq*, which tells Nyquist to play events in sequence:

```
> (defun make-sequence (duration)
   (seq (make-chord duration)
        (make-chord duration)
        (make-chord duration)))
```

The above program uses *seq* to play a sequence of three chords, as defined by *make-chord*. Since there is no provision for inputting notes in *make-chord*, the *make-sequence* will play the same chord three times.

We are now in a position to illustrate one of the most powerful features of Nyquist: the ability to apply transformations to functions as if these transformations were additional parameters. There are a number transformations available in Nyquist and they are intelligent in the sense that they were programmed with default values that often perform their tasks without the need for major adjustments. As an example of this, suppose we wish to transpose the chord sequence produced by *make-sequence* five semitones higher. In order to do this we can simply apply the transformation *transpose* to the *make-sequence* function as follows (the parameter '5' after *transpose* indicates the number of semitones for the transposition):

```
> (play (transpose 5 (make-sequence 3.5)))
```

And indeed, transformations can be applied to functions within function definitions. Example:

```
> (defun two-sequences (duration)
     (seq (transpose 5 (make-sequence duration))
          (transpose 7 (make-sequence duration))))
```

A variation on the example above is given as follows:

```
> (defun two-sequences (duration)
     (seq (transpose 5 (make-sequence duration))
          (transpose 7 (stretch 4 (make-sequence duration)))))
```

In this case, the third line specifies a nested application of transformations where *stretch* is used to enlarge the duration of the chords in the sequence and then *transpose* is used to transpose the whole sequence seven semitones upwards.

Nyquist runs under various operational systems, including Windows (95, 98, NT), MacOS and Linux/Unix. Two versions are provided on the CD-ROM: Windows and MacOS; the Linux version can be downloaded from the Carnegie-Mellon University's Web site. Installation instructions for all systems are given in the Nyquist Reference Manual.

8.3 Som-A

Som-A is a system for additive synthesis (see Chapter 3) devised by Aluizio Arcela at the University of Brasília, Brazil (Arcela, 1994). The word 'som' in Portuguese means sound and 'soma' means addition.

At the heart of *Som-A* is an elementary but smart programming language for the specification of additive instruments and scores; both are saved to a single file denominated by Aluizio Arcela as a *spectral chart*. The system provides a front-end for operation which includes an editor for writing spectral charts and facilities for playback.

The language itself is extremely simple; it only has one unit generator, called *H-unit* (Figure 8.2). An instrument is nothing more than a set of H-units, where each unit is set up to produce a partial of the desired spectrum.

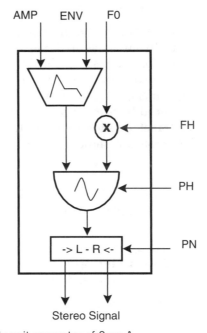

Figure 8.2 The H-unit is the only unit generator of Som-A

Each *H-unit* contains a sinusoidal oscillator, an envelope and a panning control, and each of them requires the specification of the following parameters:

- The shape of envelope and duration (ENV)
- Amplitude (AMP)
- Reference frequency; i.e. fundamental frequency (F0)
- Partial produced; in order to define the actual frequency produced (FH)
- Phase of the sinusoid in degrees (PH)
- Panning (PN)

The language has only five different instructions and its syntax is somewhat similar to Lisp; commands are lists whose first element is the instruction. Som-A instructions are *VAL*, *INS*, *EXE*, *STP* and *FIM* and all should appear in a spectral chart (see Som-A's user manual for more details):

- VAL sets initialisation values such as timing in seconds and sampling rate
- INS used to define an instrument: life-span in seconds (i.e. how long it will run), its name and a list of H-units; each unit is defined as follows (harmonic, phase, (envelope), panning)

- EXE indicates the beginning of the score section
- STP indicates the end of the score section
- FIM indicates the end of the spectral chart

The overall structure of a spectral chart is as follows:

```
(VAL <header parameters>)
(INS <instrument definition>)
(EXE <timing information>)
        (<sound event 1>)
        (<sound event 2>)
        . . .
        (<sound event n>)
(STP)
(FIM)
```

As an example, consider the following spectral chart:

```
(VAL 0 6 44100)
(INS 6 instr1
        (1 0 ((0 0) (1000 255) (0 511)) 1)
        (5 45 ((0 0) (1000 255) (0 511)) 0))
(EXE 0 6)
        (instr1 0 1 370.0 20)
        (instr1 1 1 425.3 20)
        (instr1 2 2 466.2 20)
        (instr1 4 2 622.3 20)
(STP)
(FIM)
```

The first instruction notes that the chart will produce a sound file of 6 seconds' duration at a sampling rate of 44.1 kHz. The second instruction defines one instrument, called *instr1*, whose life-span is equal to 6 seconds. The instrument produces two partials, the fundamental and the fifth harmonic. The fundamental is panned to the right channel and the fifth harmonic panned to the left. The fifth harmonic is 45 degrees out of phase in relation to the fundamental, and the envelope ((0 0) (1000 255) (0 511)) is used for both partials. The lines between instructions EXE and STP play a sequence of four pitches: 370.0 Hz, 425.3 Hz, 466.2 Hz and 622.3 Hz, respectively. The first two pitches last for 1 second each whereas the remaining two pitches last 2 seconds each.

8.4 Audio Architect

The power and flexibility of synthesis programming languages such as pcmusic and Nyquist are unquestionably impressive. However, the idea of having to write computer programs to produce sounds may not suit the appetite of some electronic musicians, particularly those used to programming sounds on analog synthesisers. But there again, nothing ventured, nothing gained – some level of computer programming will always be required if one wishes to create non-clichéd, personalised timbres on a computer.

The Karnataka Group, a software company based in London, got in at the ground floor of software synthesis programming and designed Audio Architect: a system that combines computer programming with analog-style synthesis. Audio Architect is a synthesis programming system for PC-compatible computers provided with a Microsoft Windows-based graphic interface for building instruments. Like pcmusic and Nyquist, Audio Architect is also rooted in the concept of unit generators, referred to here as *synthesis modules*. It provides a menu of synthesis modules that mimic the various components of an analog synthesiser, such as oscillators, LFOs, ADSR envelopes, filters and the like (Figure 8.3); see Russ (1996) for a discussion about analog synthesisers.

Figure 8.3 Audio Architect provides a menu of synthesis modules that mimic the various components of an analog synthesiser

An instrument, or network in Audio Architect parlance, is simply assembled by dragging synthesis modules from the menu into a working area (Figure 8.4). The connections between the modules are made simply by placing the cursor over the one module, holding the right mouse button down and dragging the wire to the next module. These connections carry *event* or *audio* signals from one module to another. The main difference between an event signal and an audio signal is that the former is used exclusively to carry control messages and not audio samples; the latter demands far more processing power for computation. Specific parameter settings for each module of the network are specified in a dialogue box which is accessed by double-clicking with the left mouse button over the module (Figure 8.5).

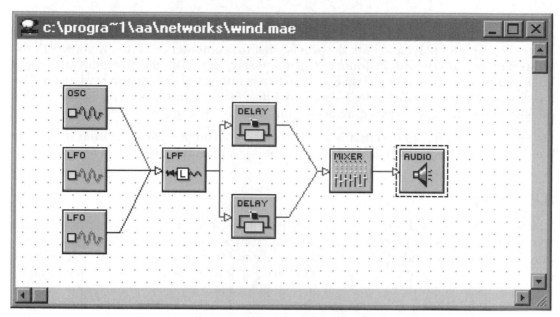

Figure 8.4 Synthesis modules are dragged from the menu into a working area and the connections between them are made by placing the cursor over one module and dragging the wire to the next module

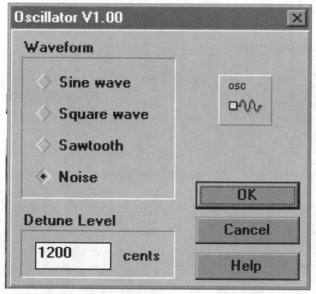

Figure 8.5 The dialogue box to set the parameters for an oscillator. Specific parameter settings for each module are specified via dialogue boxes which are accessed by double-clicking with the left mouse button over the modules

Audio Architect is a thoroughgoing example of an attempt to turn a home computer into a musical instrument. Like synthesis programming languages, Audio Architect also generates audio files for later playback. With the proviso that the computer can actually cope with the complexity of the synthesis instrument at hand, the system offers the possibility of using a MIDI controller to play it in real time as well. Furthermore, the system also reads score files for playing instruments and indeed, Audio Architect features an analog-style sequencer that can step through up to sixteen different synthesis values at a user-specified rate.

On the CD-ROM, Audio Architect is accompanied by a comprehensive tutorial on how to build your own instruments, prepared by Kenny McAlpine at Glasgow University.

8.5 Reaktor

Reaktor, by Berlin-based Native Instruments, is a complete sound design studio with almost endless possibilities for producing and performing music. Reaktor runs stand-alone or as a plug-in on Windows and Macintosh applications and it can be used either as a sampler, an effects processor, a groovebox and/or, of course, as a synthesiser. Reaktor's high-quality signal processing algorithms and 32-bit floating-point audio engine guarantee sound quality that meets the highest professional demands.

Like Audio Architect, Reaktor instruments are built by combining synthesis modules. Reaktor, however, provides modules for a greater variety of synthesis tasks other than analog-inspired ones, including standard Music N-style modules. The software features more than 200 basic modules which are the source material for building synthesisers and effect-processors. The system comes with a library containing hundreds of different instruments with thousands of presets for a wide range of sounds and audio processing

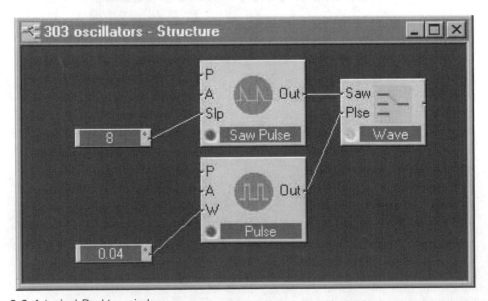

Figure 8.6 A typical Reaktor window

applications. It also features compressors, EQs, pitch shifters, filtered delays, distortion, and up to sixteen channels of surround sound.

The modus operandi of Reaktor is very similar to that of Audio Architect. Modules are selected from a menu and are interconnected in a working area to form instruments. (Figure 8.6).

From a synthesis programming point of view, one of the most important features of Reaktor is that it allows for the encapsulation of a configuration of modules into a single 'meta-module', or *macro* as it is referred to in the user manual. A macro can be saved in a file and be called up by any other instrument. This facility is very useful for building large complex instruments because it is impracticable to display too many modules in the workspace at once. Also, a whole instrument can be saved as a meta-module and be assembled together with other instruments to form an *ensemble*. Multiple polyphonic instruments can be assembled and controlled at the same time. The number and complexity of the instruments that can be put together in an ensemble depends on the processing power of the host computer.

There is so much to Reaktor that it is impossible to give a fair introduction in just a few paragraphs. The reader is highly recomended to look at Native Instrument's Web site for more information and tutorials (the Web site address can be found in the HTML document *web-refs.htm*, available in folder *various*, on the CD-ROM).

8.6 NI-Spektral Delay

NI-Spektral Delay, also by Native Instruments (the same makers as Reaktor, introduced in the previous section) uses real-time FFT (Fast Fourier Transformation) to split each channel of a stereo signal into up to 160 separately modifiable frequency bands. The level, delay time and feedback amount for each of these bands can be set separately. Additionally, various modulation effects can be applied to the signal in the frequency domain, which allows for even further sound manipulation. This software offers great possibilities for creative sound design, from subtle corrections and rich effects to the total alienation of the input signal. It is intuitive to use and all parameters are controlled from a sophisticated GUI, Graphical User Interface (Figure 8.7). It runs either stand-alone or as a plug-in. It offers professional sound quality and can be fully controlled by MIDI.

NI-Spektral Delay sports edit graphs and functions to set up the delay times and feedback levels for the bands: each of the up to 160 bands per stereo channel can be delayed separately and fed back to the delay input with an individual amount. The delay times can be set freely or according to a rhythm via a selectable note grid. The maximum delay time is 12 seconds. The effects achievable with this band delay include interesting subtle colorings as well as rhythmic courses of the partial tones and superb dense, atmospheric sound textures.

The modulation effects can be expanded by plug-ins and can be applied in the frequency domain of the audio signal, and offer many unusual processing options, including the rotation of the frequency bands and reverb simulation by 'smearing' the amplitude of the bands. The amplitudes of the bands can also be controlled by drawing in an edit graph with the mouse. The resulting filter curves can have any shape, a feature that makes all

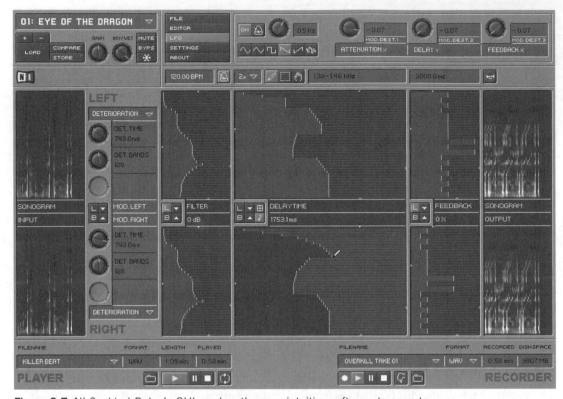

Figure 8.7 NI-Spektral Delay's GUI renders the very intuitive software to operate

imaginable settings and sweeps possible. The filter curves can be drawn commonly or separately for the two channels of a stereo signal.

All parameters such as filter, delay and feedback settings can be modulated with an integrated LFO or via MIDI. This makes it possible to create very organic and lively effects which can easily be played live. Many presets for different applications (e.g. guitar, vocals, drums, reverberations, special effects) are included in the package. It runs on Macintosh and PC-compatible platforms under Windows, either stand-alone or as a plug-in with a VST, or DirectX, compatible host program.

8.7 Praat

Praat, developed by Paul Boersma and David Weenink at the Institute of Phonetic Sciences of the University of Amsterdam, combines a number of tools into a single environment primarily aimed at supporting research in the field of phonetics, but it turns out to be an extremely useful tool for musicians as well. It is particularly well suited for those interested in working with the human voice. It includes comprehensive speech analysis and synthesis tools combined with general numerical and statistical functions, plus facilities to produce a high-quality graphic data display. Besides its comprehensive editing and analysis facilities, Praat also features a PSOLA tool (as introduced in Chapter 5), a subtractive format synthesiser with the ability to estimate the filter coefficients from given audio (Chapter 4) and the fantastic physical model of the human vocal tract discussed earlier in Chapters 4 and 6.

Boersma and Weenink present Praat as a program for handling 'objects' from the world of phonetics. An 'object' is anything that contains data that can be processed by a Praat command for doing tasks such as editing, drawing, querying, analysing, synthesising, and so on. The program features a general-purpose GUI shell consisting of two fixed windows: the *Object window*, for managing objects (i.e. data) and the *Picture window*, where the user can produce high-quality graphic displays of the objects (Figures 8.8 and 8.9). Note that the

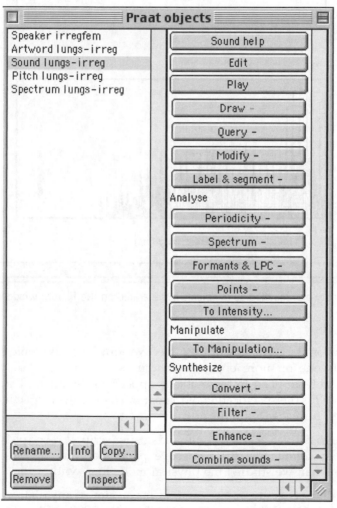

Figure 8.8 The objects are listed on the left side of the window and Praat commands on the right side. Praat automatically displays only the commands that are applicable to selected objects

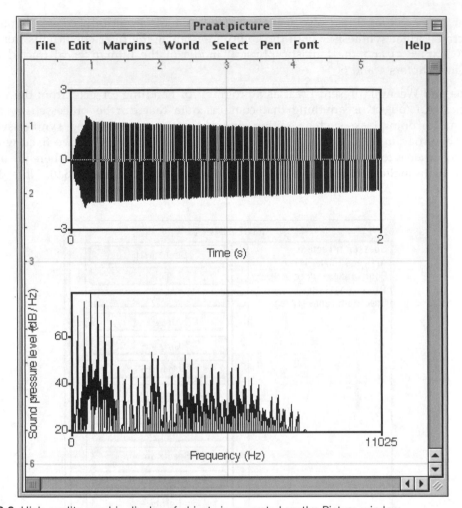

Figure 8.9 High-quality graphic display of objects is generated on the Picture window

column of buttons on the right side of the Object window is a 'dynamic menu' in the sense that if one selects one or more objects from the list on the left side, only the possible commands that can be applied to these objects appear on the menu. Each of these buttons activates a window for the specification of the parameters of the respective command; these windows are often furnished with default values.

It is also possible to activate Praat's commands using scripts. A script is a text containing sequences of Praat commands, the respective parameters and objects, as illustrated in Chapters 4 and 6 when we studied the physical modelling synthesiser.

Praat is an impressive program, it is free and it runs on almost any platform imaginable: Mac OS (PowerMac, G3, G4), Windows (95, 98, NT4, ME, 2000, XP), and all sorts of Unix machines (Linux, Solaris, SGI, Indigo, O2, Onyx, and so forth).

8.8 Reality

Reality is a software synthesiser that runs on PC-compatible computers developed by Seer Systems, Los Altos, USA. The modus operandi of Reality is not rooted in the concept of assembling instruments by interconnecting synthesis building blocks. Instead, Seer Systems ventured to implement a large multipurpose instrument that integrates different synthesis techniques. As a rough comparison, imagine Reality as the combination of an analog synthesiser, a sampler and a Yamaha VL-type of machine, integrated into one single MIDI instrument accompanied by a comprehensive patch editor. The major drawback of this approach is that Reality limits the sound designer to the synthesis techniques that it can produce, but this is the price one has to pay for high-level systems of any kind.

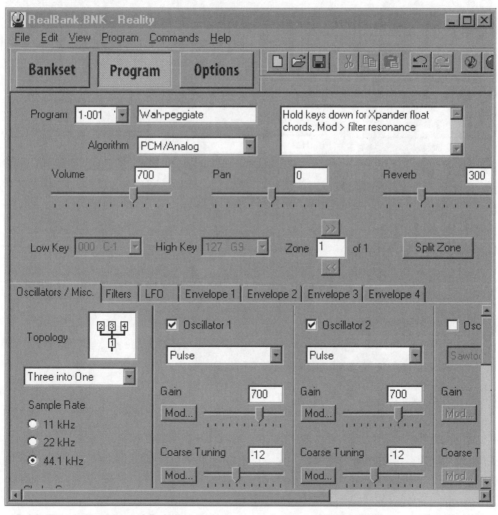

Figure 8.10 The main screen of Reality provides access to three main working areas of the system: Bankset, Program and Options

Nevertheless, Reality is easy to use and it is much quicker to learn than learning actual programming languages, such as Nyquist, or programming systems, such as Audio Architect; Reality is flexible for building its 'patches' and it can hold more than one thousand patches in its memory. Also, it offers 64-voice polyphony and is multi-timbral to sixteen MIDI-channels.

The system is structured as follows:

- At the highest level is the *bankset*: a storage structure that can hold up to 1336 programs, arranged in groups of 128 (because of the MIDI way of enumerating units, i.e. from 0 to 127)

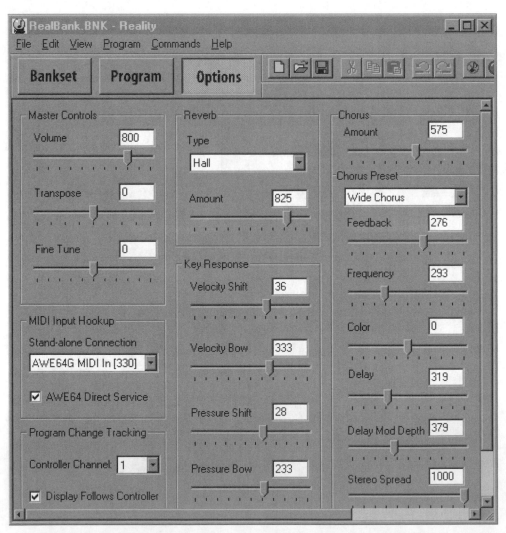

Figure 8.11 The Options view is used to set up a number of global playback tools, including the master volume, a reverb, a chorus and MIDI source

- A *program*, which may be a *patch* or a *patchwork*, is more or less equivalent to a timbre on a standard MIDI synthesiser. Programs can be selected from any MIDI controller able to send program change messages; see Rumsey (1994) for an introduction to MIDI. A patch is simply the set-up for a timbre using any of the synthesis capabilities of the system, whereas a patchwork may combine different types of timbres

Reality provides two main groups of synthesis facilities, referred to as *PCM/analog settings* and *algorithmic methods*. The former includes a number of settings for implementing synthesis techniques inspired by analog synthesisers, such as FM and additive synthesis (see Chapters 2 and 3). In this case, there are four oscillators that can be configured in a number of topologies, four envelope generators, four filters and a number of controllers that indeed resemble the functioning of an analog synthesiser. The oscillators can produce a number of different waveforms, including sawtooth, triangle, pulse and sinusoid; oscillators can also take in sound samples to act as a template waveform. As for the *algorithms methods*, the program provides settings for a number of synthesis methods inspired by source modelling approaches (see Chapter 4).

The main screen of the program provides access to the three main working areas of the system: the *Bankset* view, where one selects programs and sampled wave templates; the *Program* view (Figure 8.10), where the editable controls of a sound are located; and the *Options* view, where one can set up a number of global playback utensils such as the master volume, a reverb, a chorus and the MIDI source (Figure 8.11).

8.9 Diphone

Diphone, developed at Ircam in Paris, is a program for making transitions between sounds; an effect commonly referred to as *sound morphing*. The program runs on Macintosh computers and is part of a set of programs supplied to the subscribers of the Ircam Forum. Originally, the term 'diphone' refers to a transition between two phonemes, but for musical purposes this notion has been expanded to mean the transition between any two sounds, not necessarily of vocal origin.

Ircam's Diphone addresses one of the most challenging synthesis problems of both artificial speech and sound composition: the concatenation of sound sequences. In computer-synthesised speech the algorithm used to utter a specific syllable, such as 'mu', may not produce satisfactory results on different words. For instance, take the words 'music' and 'mutation'. The computer's rendition of the syllable 'mu', for the word 'music', used at the beginning of the word 'mutation' sounds artificial because the transition (i.e. the *diphone*) between 'u' and 's' in the word 'music' and the transition between 'u' and 't' in the word 'mutation' involve different spectral behaviour. This transition problem is also evident in music composition where certain articulations and musical passages are clearly more appropriate to our ears than others. This problem is even more challenging for electronic musicians because they deal with a much larger repertoire of sounds to combine and articulate.

Although this book does not focus on music composition, Diphone is introduced here as an example of a synthesis system that was primarily designed to address a problem pertinent to all electronic musicians, and more specifically to those working with sampled sounds and sound montage; e.g. *acousmatic music* (Bayle, 1993; Norman, 1996). Composers working with recorded sound normally concatenate the sounds using splicing or crossfading techniques. In most cases, however, the results do not sound satisfactory because the inner contents of the sounds involved do not always match. Diphone attempts to render this task more effective by using analysis and resynthesis techniques (refer to Chapter 3).

Each sound sample in Diphone is submitted to an analysis stage that extracts crucial information on how its spectrum evolves in time. This analysis provides a 'multi-parametric' representation of the sound; it contains information about its fundamental frequency (i.e. pitch), plus the frequency, the amplitude and the phase of each component of its spectrum. The advantage of such representation over straight sampling is that this information can be manipulated individually and mapped to the parameters of a synthesis algorithm suited for resynthesising the sound; for example, *additive synthesis* (discussed in Chapter 3). If the analysis information is not changed, then the outcome of the resynthesis process should be equal to the original sound. Conversely, if the analysis information is changed, then the resulting resynthesis will sound different. For example, if the fundamental frequency is changed, then the result will be a sound at a different pitch. Note, however, that this modification is by no means equivalent to playing back the sound at a faster or slower speed, as samplers do. In this case, the timbre may be considerably changed, whereas in Diphone the timbre of the sound would remain intact.

Each sound in Diphone is referred to as a *segment*. The program produces a set of analysis data for each segment and stores them on a *dictionary* of segments. Diphone provides an intuitive user interface for operation, in which one can drag segments from a dictionary and drop them onto a working area, referred to as the *sequence window* (Figure 8.12). At the top of the sequence window the computer displays the segments for concatenation and at the bottom it provides a menu of 'parameters' for monitoring the concatenation.

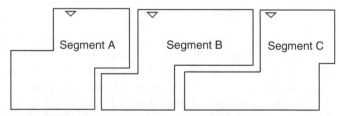

Figure 8.12 Diphone concatenates the sounds by applying an algorithm that interpolates the analysis data of neighbouring segments

The program concatenates the sounds by applying an algorithm that interpolates the analysis data of neighbouring segments. It is important to note that Diphone does not manipulate the sounds directly, only the analysis data. The whole sequence is synthesised only when the user commands the computer to do so.

Each segment is represented by an icon and it has three distinct areas: a central area and two adjacent interpolating areas (Figure 8.13). The user can adjust the extent of the

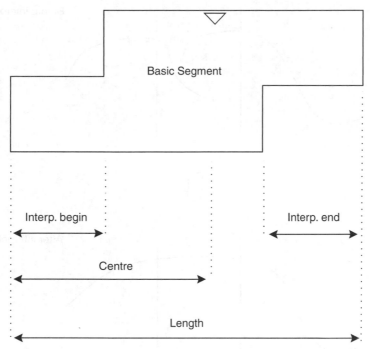

Figure 8.13 Each segment is represented by an icon and it has three distinct areas: a central area and two adjacent interpolating areas

interpolating areas and define the type of interpolation algorithm for each neighbourhood; e.g. linear, logarithmic, etc.

One of the major strengths of Diphone is the possibility to edit and monitor the concatenation graphically. The bottom area of the sequence window lists the parameters to be interpolated and other global settings, such as a transposition index and a gain factor for amplification or attenuation of the signal. Here one can pull down graphic breakpoint functions to monitor the behaviour of these parameters, with options to visualise the plot before and after the interpolation takes place (Figure 8.14). Furthermore, there are facilities for editing the plot manually with a 'pencil'.

Although originally designed to work with additive synthesis, Diphone also supports other synthesis techniques such as formant synthesis (discussed in Chapter 3) and source modelling (Chapter 4).

8.10 The CDP sound transformation toolkit

The CDP sound transformation toolkit is part of a large package called *The CDP Computer Music Workstation* currently available for the Atari Falcon, PC-compatible and Silicon Graphics platforms. CDP stands for Composer's Desktop Project and it was created in the mid-1980s in York by a pool of British composers and sound engineers with the objective

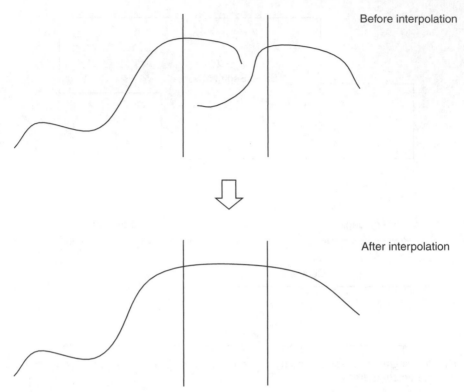

Figure 8.14 Concatenation parameters can be monitored graphically before and after the actual interpolation takes place

of producing affordable computer systems for music composition. At a time when computer music equipment could be sustained only by wealthy institutions, the CDP team managed to manufacture a fairly powerful system that worked on Atari ST computers provided with DAC/ADC hardware manufactured in-house and a PCM converter. The CDP system soon became very popular among composers and educators in Britain and elsewhere.

The CDP package is formed by a number of individual programs, each dedicated to a specific task. The package is subdivided into three groups of programs: (1) Synthesis, (2) Groucho Signal Processing and (3) Spectral Sound Transformation. The first group includes Csound, a Music N-style synthesis programming language that resembles pcmusic in many ways, but the last two groups constitute CDP's most powerful and unique pieces of software.

The Groucho Sound Processing group includes programs for manipulating the samples of a sound file in the time-domain. There are more than 100 different programs in this group, which allow for many ways to manipulate a sound, including all the time-based approaches described in Chapter 5.

As for the Spectral Sound Transformation group, it includes dozens of highly specialised programs for spectral manipulation, which constitute one of the most powerful toolkits of its kind. They deal directly with the timbre or tone colour of a sound, thus making possible the modification of timbre in unprecedented ways. The Spectral Sound Transformation tools carry out a spectral analysis of the sound to user-defined degrees of complexity, whereupon the numerical data can be manipulated in many different ways (see Chapter 3).

8.11 Sound Shaper

Sound Shaper is a GUI for the CDP sound transformation toolkit, developed by Robert Fraser. This comprehensive piece of work is written in Delphi and is available for the IBM PC-compatible Windows platforms. It 'sits on top of' and transparently makes use of the standard CDP command line set of programs, so it will work immediately with any current Release 4 version of the CDP System.

One of the key features of Sound Shaper is its practical and concise layout. As indicated by Acher Hendirch on the CDP Web site (see *web-refs.htm* file in the folder *various*), the fundamental sequence of operations in a sound design package such as CDP is 'hear–alter–hear again' and Sound Shaper fully supports this basic tripartite sequence by making its main window a play mechanism, with the full functionality of the CDP system available above it in the form of drop down menus (Figure 8.15).

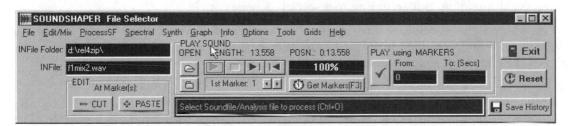

Figure 8.15 The main window of Sound Shaper was designed to foster the typical 'hear–alter–hear again' sound design sequence

When a process is completed, the output soundfile appears in this main window, ready to be played. Thus one can hear the source sound, go to one of the CDP functions to alter it, and then hear the result, all from the main window of the GUI. Furthermore, this window is very compact, enabling the composer to place another application, such as a soundfile editor, on screen at the same time.

Sound Shaper features a straightforward breakpoint editor for creating time-varying contours; remember that many parameters of the CDP System support time-varying parameters and this is an essential requirement for designing dynamically rich sounds. This facility is easily accessed from the drop-down menu of the main window.

Sound Shaper also features a useful *History Function* for building up libraries of complex functions which can be run at a stroke. When activated, Sound Shaper will record all the processes you have run by writing a text file of the respective command lines. This text file can then be edited and used in MS-DOS as a batch file with different soundfile input(s). Experienced composers will gradually create their own set of operations central to their own composing style. Moving between Sound Shaper and MS-DOS is completely transparent because the interface runs with the standard CDP programs.

The demo version on the CD-ROM has limited functionality, but shows the full menu structure and layout of the Sound Shaper interface. Of special interest on the accompanying CD-ROM is the Sound Shaper tutorial specially prepared for this book by Robert Thompson (in the *sndshap* folder).

8.12 LASy

LASy (Linear Automata Synthesis) is a program made specifically for cellular automata lookup table synthesis. It was designed by Jacques Chareyron (1990), at the University of Milan's 'Laboratorio di Informatica Musicale' (LIM), Italy. LASy runs on Macintosh computers and it is part of a pool of programs developed at LIM, which together form a larger integrated computer music workbench called *Intelligent Music Workstation*, or *IMW*.

The cellular automata lookup table synthesis technique is discussed in Chapter 4. LASy basically works by applying a cellular automaton to a lookup table containing an initial waveform. At each playback cycle of the lookup table, the cellular automaton algorithm processes the waveform. The intention is to let the samples of the lookup table be in perpetual mutation, but according to a sort of 'genetic code'.

Considering that LASy's synthesis technique does not demand heavy computation to process the samples, sounds can be produced in real time via a MIDI controller, mouse or even the computer's alphanumeric keyboard. The program provides facilities for configuring the MIDI communication.

LASy instruments are defined in terms of an initial waveform, a rule for the cellular automaton (referred to as the *transition rule*) and envelopes for amplitude and pitch. The system provides tools for the individual specification of these components so that the user can build their own library of waveforms, rules and envelopes (Figure 8.16). Instruments are either created by combining components selected from these libraries or from scratch on a window, where the user can set up all these parameters at once (Figure 8.17).

Initial waveforms are created within the system itself either by using a breakpoint function or by adding sinusoids, but sampled sounds may also be used. A great variety of instruments can be created by combining different waveforms and transition rules.

LASy is able to synthesise a large variety of sounds with diverse spectral evolutions, particularly sounds with fast transients at the very beginning of the sound. The program is particularly good for producing wind-like and plucked strings-like sounds. Yet, the ingredient that still makes LASy unique is its ability to synthesise unusual sounds but with some resemblance to the real acoustic world.

Jacques Chareyron classifies the output of LASy into three main groups, according to the type of cellular automata rules employed (Chareyron, 1990):

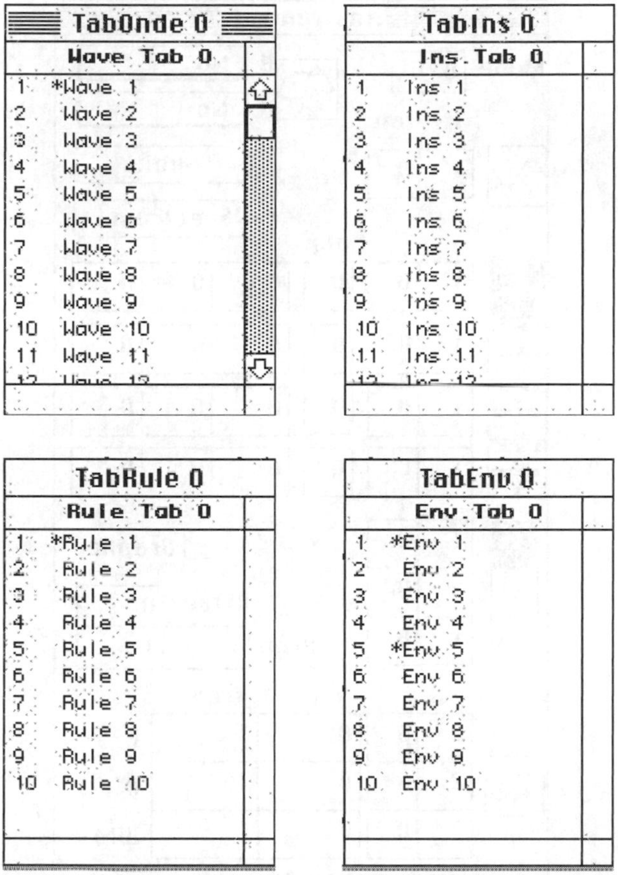

TabOnde 0			TabIns 0	
Wave Tab 0			**Ins.Tab 0**	
1 *Wave 1	⇧		1 Ins 1	
2 Wave 2	▮		2 Ins 2	
3 Wave 3			3 Ins 3	
4 Wave 4			4 Ins 4	
5 Wave 5			5 Ins 5	
6 Wave 6			6 Ins 6	
7 Wave 7			7 Ins 7	
8 Wave 8			8 Ins 8	
9 Wave 9			9 Ins 9	
10 Wave 10			10 Ins 10	
11 Wave 11	⇩		11 Ins 11	
12 Wave 12			12 Ins 12	

TabRule 0		TabEnv 0	
Rule Tab 0		**Env Tab 0**	
1 *Rule 1		1 *Env 1	
2 Rule 2		2 Env 2	
3 Rule 3		3 Env 3	
4 Rule 4		4 Env 4	
5 Rule 5		5 *Env 5	
6 Rule 6		6 Env 6	
7 Rule 7		7 Env 7	
8 Rule 8		8 Env 8	
9 Rule 9		9 Env 9	
10 Rule 10		10 Env 10	

Figure 8.16 LASy instruments are defined in terms of an initial waveform, a rule for the cellular automaton and envelopes for amplitude and pitch

- *Sounds with simple evolution leading to a steady-state ending*: simple cellular automata rules (refer to the system documentation) generate monotonous evolution of the sound spectrum, where the spectral envelope follows either an increasing or decreasing curve, leading to a steady-state ending
- *Sounds with simple evolution but with no ending*: by increasing the complexity of the cellular automata rules one obtains endless successions of similar but not completely identical waveforms
- *Everlasting complex sounds*: the more complex the rules, the more unpredictable is the behaviour of the cellular automata, and therefore the more complex the evolution of the sound

This classification is, of course, very general inasmuch as the boundaries between the categories are crudely vague. Nevertheless, as LASy's author himself suggests, they are a good starting point for further experimentation.

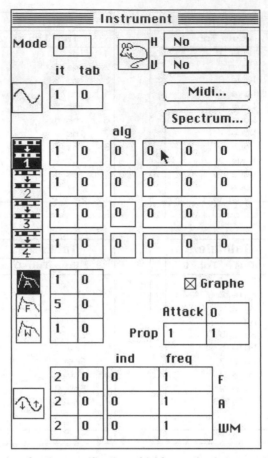

Figure 8.17 The main window for the specification of LASy synthesis parameters

8.13 Wigout and TrikTraks

Wigout and TrikTraks form a system for PC-compatible platforms (MS-DOS and Windows) specifically designed for sequential waveform composition (see Chapter 5). The system was designed by the composer Arun Chandra in collaboration with researchers of the School of Music at the University of Illinois, USA (Chandra, 1994).

Like Som-A, the heart of both Wigout and TrikTraks is a pair elementary programming languages for the specification of waveforms and transitions. Each of them has its own sound compiler.

Wigout works by generating sequences of sample segments that, taken together, compose a waveform. It can generate complex combinations of waveforms of arbitrary length, limited only by the amount of memory on the computer (see Chapter 5 for more details about the sequential waveform composition technique).

In Wigout, a sound is specified in terms of three separate files referred to as *segments*, *states* and *events*. The *segments* file contains descriptions for the segments to be used to compose the waveform. Each segment is described in terms of amplitude and duration (measured in samples). In addition, a segment also has a range (maximum and minimum values) and an increment of change. It can have three different shapes: square, triangular or curved, referred to as a *wiggle*, a *twiggle*, or a *ciggle*, respectively. A segment is defined by three lines of code. For example (comments are always preceded by the symbol '#'):

```
a1 wiggle              # identifier, type
10 20 1 1              # initial, max, min, inc (duration)
 0 32000 -32000 4000   # initial, max, min, inc (amplitude)
```

The first line declares a wiggle segment labelled as *a1*. Then, the second line determines the duration of the wiggle. The first item on this line is its initial duration, followed by its maximum duration, minimum duration, and increment. All values are given in samples. In this case, the initial duration is ten samples, the maximum duration it can reach is twenty samples, and the minimum duration it will reach is one sample. On each iteration of this segment, the duration will change by one sample. Finally, the third line describes the amplitude of *a1*: its initial value, followed by its maximum value, minimum value, and increment. In this case, the initial amplitude will be 0 but it will increase to 32 000 and decrease to –32 000, and will change by 4000 at each iteration. Every time this segment is iterated, its duration and its amplitude will increase by their respective increments. Once they reach their maximum ranges, they will start shrinking and when they reach their minima, they will start increasing again.

The *states* file contains a specification of sequences of segments that make up a particular combination, or state, for generating waveforms. For example:

```
# id segments
s1 a1 a2 a3
s2 a2 a3 a2 a3 a4
```

In this case there are two states, labelled as *s1* and *s2*, respectively. The first state consists of three segments: *a1* followed by *a2*, followed by *a3*, and the second state consists of five segments: *a2* followed by *a3*, followed by *a2* again, followed by *a3* again, followed by *a4*. A state is iterated from its start time until its end time, according to the specifications given in the events file. At every iteration, each of its segments changes its duration and its amplitude.

The *events* file contains a list of states, their start and end times (in seconds), and their stereo location (a number between 0 and 1):

```
s1 0 5 0  # state label, start time, end time, stereo location
s2 0 5 1
```

In this example, state *s1* (as defined in the states file) begins at time 0 seconds, iterates until time 5 seconds, and will be played on the left of the stereo field (left = 0). Similarly, state *s2* also begins at 0 seconds and iterates until 5 seconds, but it will be played on the right of the stereo field (right = 1).

TrikTraks complements Wigout by providing the means for the specification of *transformational paths* during the production of a waveform. Here the electronic musician can specify a number of different functions to change the waveform, including sine, triangular and polynomial trajectories.

8.14 SMS

SMS is an analysis and resynthesis program (see Chapter 3) designed by Xavier Serra, of the Pompeu Fabra University in Barcelona, Spain. The program on the accompanying CD-ROM runs on PC-compatible computers under Windows and it has been developed as a graphic front-end (Figure 8.18) for a powerful analysis and resynthesis consort, originally available

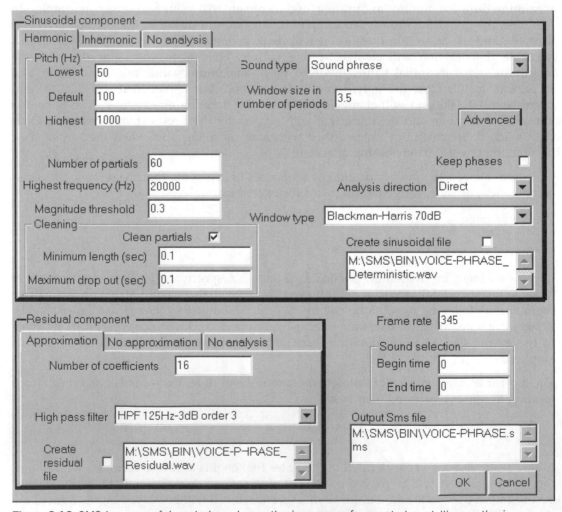

Figure 8.18 SMS is a powerful analysis and resynthesis program for spectral modelling synthesis

as a set of individual programs that were activated manually via DOS or Unix command lines.

The new Windows-based front-end integrates the whole SMS consort into one program and includes a number of additional tools for the display and manipulation of the analysis data, as well as graphic facilities for the specification of resynthesis parameters (Figures 8.19 and 8.20). The program offers gadgets for three main types of tasks: *analysis, transformation-resynthesis* and *event-list control*. Whilst the analysis group involves tools for displaying a

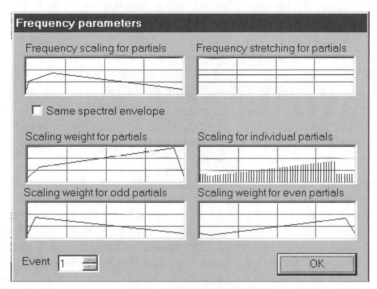

Figure 8.19 SMS includes a number of tools for the display and manipulation of analysis data

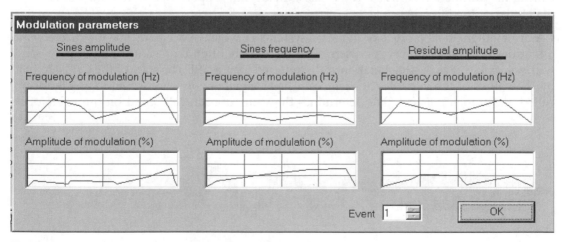

Figure 8.20 The Modulation parameter window allows for the specification of modulation parameters for resynthesis

sound and the analysis data generated by the system, the transformation-resynthesis group includes facilities for manipulating the analysis data for resynthesis. Finally, the event-list control tools allow for detailed control of resynthesis whereby the musician can define tasks such as spectral interpolation of different sounds.

Although SMS is an abbreviation for Spectral Modelling Synthesis, it is important to bear in mind that in this book the author uses the term 'spectral modelling' to refer to a class of synthesis techniques and not to a specific program. The SMS program on the CD-ROM in fact embodies the *resynthesis by reintegration of discarded components* technique described in Chapter 3. For an in-depth discussion on the inner functioning of SMS the reader should consult the documentation provided with the program.

8.15 Chaosynth

Chaosynth is a granular synthesis program (see Figure 8.23) that uses cellular automata to control the production of the sounds; granular synthesis is discussed in Chapter 5. This program is a version for Macintosh and IBM PC-compatible under Windows computers, implemented by Joe Wright, at Nyr Sound, from an earlier version designed by this author in the early 1990s.

The granular synthesis of sounds involves the production of thousands of short sonic particles that are combined to form larger complex sound events (Figure 8.21). This synthesis technique is inspired by the British physicist Dennis Gabor's famous proposition that large, complex sound events are composed of streams of simple acoustic particles (Gabor, 1947). Norbert Wiener, one of the pioneers of Cybernetics, also adopted a similar concept to measure the information content of a sonic message (Wiener, 1964). It was the composer Iannis Xenakis (1971), however, who suggested one of the first applications of granular sound representation for sound composition purposes. Since then, a few granular synthesis systems have been proposed but, so far, most of these systems use stochastic methods to control the production of the sonic particles. Chaosynth proposes a different method: the use of cellular automata (CA).

CA are mathematical models of dynamic systems in which space and time are discrete and quantities take on a finite set of discrete values. CA are often represented as a regular array with a variable at each site, metaphorically referred to as a cell. The state of the CA is defined by the values of the variables at each cell. The automata evolve according to an algorithm, called a *transition function*, that determines the value of each cell based on the value of its

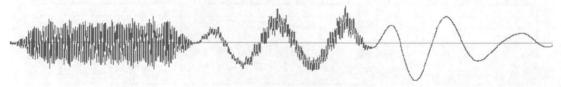

Figure 8.21 Granular synthesis of sounds involves the production of thousands of short sonic particles that are combined to form larger, complex sound events

neighbourhood. As implemented on a computer, the cells are represented as a grid of tiny rectangles whose values are indicated by different colours (Figure 8.22).

A wide variety of CA and transition functions have been invented and adapted for many modelling purposes in many scientific areas, including physics, computing, biology and meteorology. CA have also attracted the interest of musicians because of their organisational principles. Various composers and researchers have used CA to aid the control of both higher-level musical structures or musical form (Miranda, 1993), and lower-level structures, such as the spectra of individual sound events. Chaosynth uses CA to control the lower-level structure of sounds.

A demonstration version of the Chaosynth (for Macintosh and PC) plus more information about its inner functioning, as well as comprehensive system documentation and examples, can be found (in the folder *nyrsound*) on the accompanying CD-ROM.

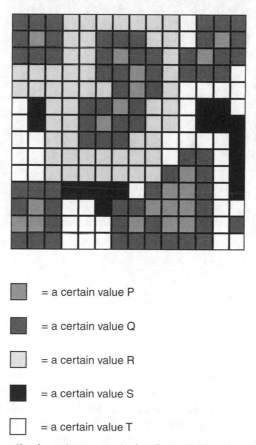

☐ = a certain value P

☐ = a certain value Q

☐ = a certain value R

☐ = a certain value S

☐ = a certain value T

Figure 8.22 Most granular synthesis systems use stochastic methods to control the production of the sonic particles, but Chaosynth uses cellular automata. The state of the cellular automata is defined by the values of the variables at each cell. As implemented on a computer, the cells are represented as a grid of tiny rectangles whose values are indicated by different colours

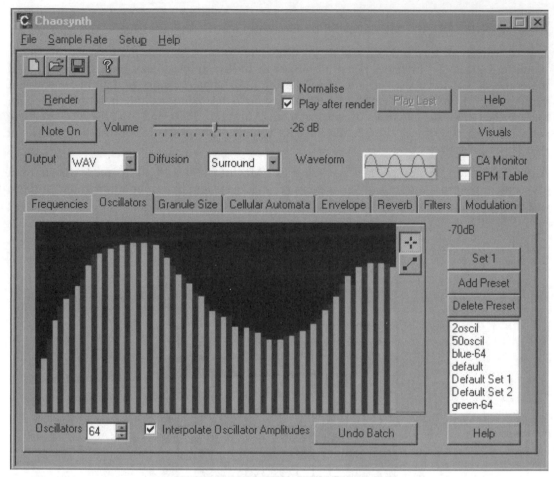

Figure 8.23 Chaosynth is an innovative granular synthesiser that used cellular automata to control the productions of the grains

8.16 Pulsar Generator

Pulsar Generator (or simply PG) designed by Alberto de Campo and Curtis Roads, implements the Pulsar synthesis technique discussed in Chapter 5; it runs on Macintosh computers.

PG's interface is neat and allows for straightforward control of the various parameters of Pulsar synthesis (Figure 8.24). The user is presented with a collection of independent windows, most of which display wavetables and envelope controls for producing the pulsar trains. The contents of these wavetables can be edited simply by Command-clicking and dragging the breakpoints. The buttons E, S, L, etc. on the side of each wavetable window give access to tools for a number of tasks such as the ability to rescale the range of the waveform within the window, loading and saving the content of the window in a file, and so on.

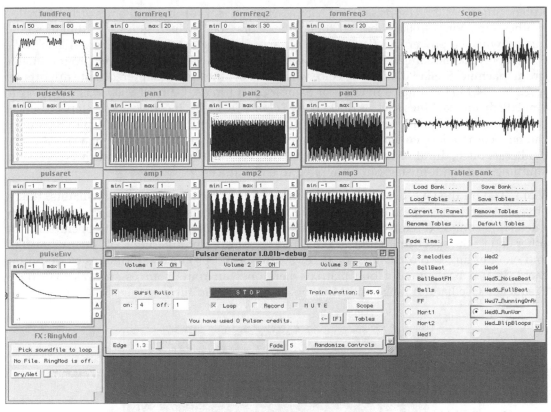

Figure 8.24 Pulsar Generator presents itself to the user with a very neat interface, which allows for immediate exploration of the potential of the Pulsar technique

The *Control Panel* lets the user start and stop a pulsar train. The duration of a pulsar train is set in the box on the right side labelled *Train Duration* and it can be set to repeat itself if the *Loop* option is checked. By clicking on the *Record* button one can save a pulsar train in a soundfile but what is interesting here is that the user can change the sound in real-time, while the it is being saved. The recorded file goes into the PG folder under the name *PulsarGenRec*.

A quick-start overview of the system is given in the document *PG ManualWord98* that is available within the PG materials. Fairly informative documentation is also provided via the program's own Help facility, including a complementary introduction to the Pulsar synthesis technique.

8.17 Koblo Vibra 1000

Vibra 1000 is monophonic stereo synthesizer for Macintosh made by Koblo, in Denmark. In fact Vibra 1000 is the baby of an impressive family of synthesisers, the Koblo Studio 9000. This family comprises five synthesisers in one versatile bundle. These synthesisers can perform in real time and seamlessly integrate into any digital audio workstation.

Koblo Studio 9000 comprises the Vibra series, a set of monophonic virtual analog synthesizers: Vibra 1000, Vibra 6000 and the Vibra 9000, which gives you digital sound quality with the power and flexibility of advanced analog synthesisers. There is also Stella 9000, a polyphonic sample-based synthesiser and Gamma 9000 a polyphonic programmable drum machine. Stella 9000 is an 8-voice wavetable synthesiser (Chapter 2) that supports up to 32-bit, 44.1 kHz samples in AIFF or SampleCell formats. Gamma 9000 was designed to emulate drum machines such as the famous Roland TR series. It comes with a step sequencer conataining a row of 16 step-entry buttons for creating rhythmic patterns. Up to 8 patterns can be recorded, looped and assembled into 'songs'. One can easily highlight a specific beat or alter its pitch, duration and loudness (i.e. MIDI velocity) and operations such as reversing, shuffling and randomising can be applied to entire tracks.

Vibra 1000 on the CD-ROM is fully operational and it comes with basic oscillators, filters, envelopes and the top-quality 32-bit stereo DSP algorithms enabling the creation of very interesting sounds (Figure 8.25). It is an extremely easy piece of software to operate. At the other end of the spectrum, Vibra 9000 has more sophisticated filtering tools than Vibra 1000 and one can link two oscillators for stereo synthesis and produce 'fat' timbres by de-tuning them. It also has a modulation matrix where one can route sources such as LFO (Low Frequency Oscillators) range and MIDI note-on velocity to filter settings.

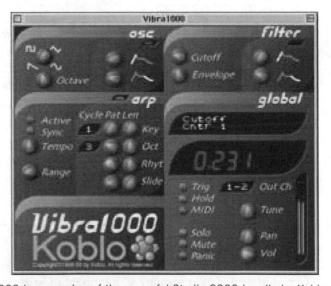

Figure 8.25 Vibra 1000 is a member of the powerful Studio 9000 bundle by Koblo

8.18 crusherX-Live

'crusherX-Live' is a shareware program for MS Windows designed by Joerg Stelkens, in Munich, Germany. It implements a time-based synthesis engine referred to by Stelkens as 'vapor synthesis'. In essence it is an enhanced implementation of the granular sampling approach to granular synthesis discussed in Chapter 5. It uses either internal oscillators, recorded samples or real-time stereo input to produce sounds. The program features a

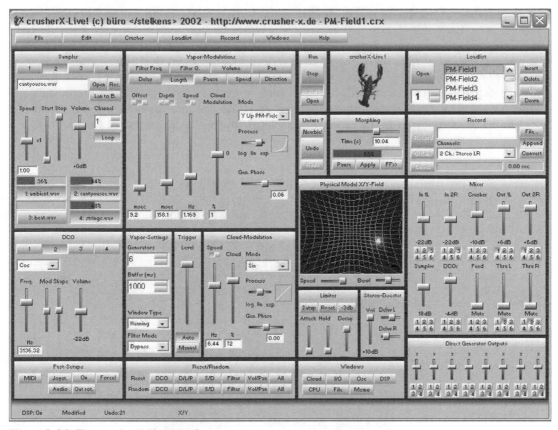

Figure 8.26 The crusherX-Live interface

multi-dimensional morphing engine that allows for the creation of clouds of sweeping and shifting sounds. The morphing time can range from 1 millisecond up to 1 hour.

Basically the program works as follows (Figure 8.26). On the left side of the GUI, the user specifies the source signals that will be 'crushed' (i.e. granulated). These can be recorded samples or signals generated by the program's own internal oscillators. Optionally, these signals can be mixed with a live-input signal in the mixer, on the right side of the GUI. The sum of these signals is then routed to the 'vapor engine', whose controls are available in the middle section of the GUI. Here the user can control how these signals will be granulated and add several other modulation effects. Nearly all faders and knobs of the GUI can be controlled via a MIDI controller or joystick in real time.

8.19 Virtual Waves

Virtual Waves, by Synoptic, is a synthesis programming system for PC-compatible computers which includes not only elementary *sound synthesis* tools but also *sound processing* and *sound analysis* facilities.

The user-friendly graphic interface of Virtual Waves greatly facilitates the straightforward creation of instruments (Figure 8.27). Once the synthesis parameters for each module have been adequately adjusted, the sound is then computed. Although Virtual Waves can be very fast for computing simple instruments, the system has not been designed to be 'played' in real time. Also there is no provision for playing back note lists or scores. Virtual Waves should primarily be used as a sound design tool and not as a musical instrument; once the sound has been created, it can either be saved onto a file for later use in some other musical application or transferred to an external sampler or to the RAM of a sound card.

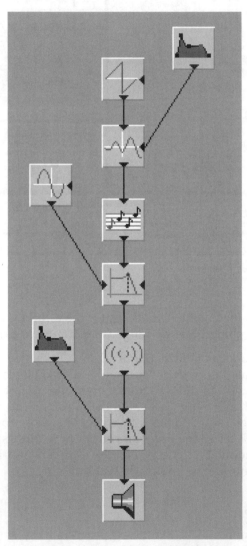

Figure 8.27 The friendly graphic interface of Virtual Waves facilitates the straightforward creation of instruments, which are produced by interconnecting unit generators

The basic building blocks of Virtual Waves for making instruments are called *modules*. The system provides a number of highly advanced modules, some of which implement fairly sophisticated synthesis techniques. For example, the system includes a module which alone is almost equivalent to the Yamaha DX7 synthesiser: it contains a six-operator FM synthesis architecture with thirty-two preset algorithms.

Since there is no efficient provision for playing an instrument via a controller or a score file, the parameters for each module must previously be set beforehand. Each module is provided with default parameter values which will be implemented if the user does not choose to edit them. Such parameters are edited via the *Edit window*, which is opened by double-clicking on the module's icon in the workspace. Figure 8.28 illustrates the *Edit window* for the *Oscillator* module. This module generates a periodic signal using different waveforms chosen from a menu; the signal can be a sinewave, a square wave, a sawtooth wave, a triangular wave or a pulse wave.

The complete list of modules is located to the left of the workspace and a module is selected simply by clicking on its name. When the mouse is moved over the workspace area an icon for the selected module automatically appears, and it can be placed anywhere within the area. The modules of the system are divided into three categories: *Generators*, *Processes* and *Analyses*.

Generators are modules which create different types of signals, which are either *audio signals* (e.g. oscillator and noise generator) or *control signals* (e.g. envelope generator). Control signals are used to control the synthesis parameters of other modules; for example, the cut-off frequency of a resonant filter. There is a variety of generators: some of them are basic synthesis units (e.g. oscillator and envelope) but most of them encompass an entire synthesis algorithm (e.g. FM, FOF and Karplus–Strong algorithm).

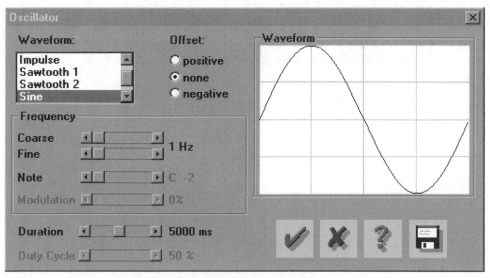

Figure 8.28 The Edit window for the Oscillator module. Each module is provided with default parameter values which can be edited via its Edit window

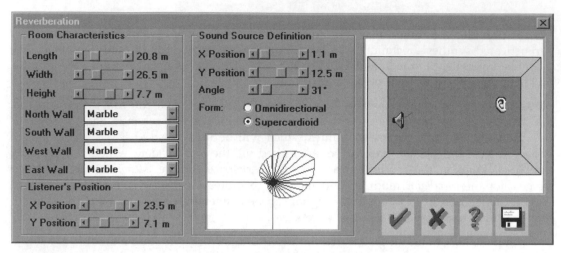

Figure 8.29 The Reverberation module is able to model the characteristics of various room acoustics, such as the dimensions of the room and absorption properties of the wall materials

The *Processes* category includes a number of modules to modify specific attributes of a given sound. Processes are by their very nature required to have at least one input and an output, with the exception of the *Output* module, which is always the last component of an instrument. This category adds all the capabilities of a powerful sound editor and sampling system to the synthesis facilities of the system. Indeed, it embraces a number of signal processors and effect units, including a powerful reverb module with the ability to model the characteristics of various room acoustics, such as the dimensions of the room and absorption properties of the wall materials (Figure 8.29).

The modules of the *Analysis* category analyse the signal passed through them without changing it in any way. Their purpose is solely to generate a visual representation of the analysis data. The available analysis techniques include STFT (see Chapter 3) and sonogram.

APPENDIX 1 Mathematical specifications

Amplitude modulation

The amplitude of the modulator is specified by an amount of the offset amplitude value (a_c) in relation to the modulation index (mi): $a_m = a_c \times mi$.

In simple AM, the spectrum of the resulting signal contains energy at three frequencies: the frequency of the carrier (f_c) plus two sidebands ($f_c - f_m$ and $f_c + f_m$ respectively). The amplitude of the carrier frequency remains unchanged, whilst the amplitudes of the sidebands are calculated as follows: $a_c \times (0.5 \times mi)$.

Frequency modulation

List of scaling factors (not exhaustive)

i	$B_0(i)$	$B_1(i)$	$B_2(i)$	$B_3(i)$	$B_4(i)$	$B_5(i)$	$B_6(i)$	$B_7(i)$	$B_8(i)$
0.0	1.000	0.000	0.000	0.000	0.000	0.000	0.000	0.000	0.000
0.5	0.938	0.242	0.030	0.002	0.000	0.000	0.000	0.000	0.000
1.0	0.765	0.440	0.115	0.019	0.002	0.000	0.000	0.000	0.000
1.5	0.512	0.558	0.232	0.060	0.011	0.001	0.000	0.000	0.000
2.0	0.223	0.576	0.352	0.129	0.034	0.007	0.001	0.000	0.000
2.5	−0.048	0.500	0.446	0.216	0.073	0.020	0.004	0.000	0.000
3.0	−0.260	0.340	0.486	0.309	0.132	0.043	0.011	0.002	0.000
3.5	−0.380	0.137	0.458	0.386	0.204	0.080	0.025	0.006	0.001
4.0	−0.400	−0.066	0.364	0.430	0.281	0.132	0.050	0.015	0.004
4.5	−0.032	−0.231	0.217	0.424	0.348	0.194	0.084	0.030	0.009
5.0	−0.177	−0.327	0.046	0.364	0.391	0.261	0.131	0.053	0.018
5.5	−0.006	−0.341	−0.117	0.256	0.396	0.320	0.186	0.086	0.033
6.0	0.150	−0.276	−0.242	0.115	0.357	0.362	0.245	0.130	0.056
6.5	0.260	−0.153	−0.307	−0.035	0.274	0.373	0.300	0.180	0.088
7.0	0.300	−0.004	−0.301	−0.167	0.157	0.347	0.340	0.233	0.128
7.5	0.266	0.135	−0.230	−0.258	0.023	0.283	0.354	0.283	0.174
8.0	0.171	0.234	−0.113	−0.291	−0.105	0.185	0.337	0.320	0.223
8.5	0.041	0.273	0.022	−0.262	−0.207	0.067	0.287	0.337	0.270
9.0	−0.090	0.245	0.145	−0.180	−0.265	−0.055	0.204	0.327	0.305
9.5	−0.194	0.161	0.228	−0.065	−0.270	−0.161	0.100	0.286	0.323
10.0	−0.245	0.043	0.254	0.058	0.220	−0.234	−0.014	0.216	0.317

Single carrier with parallel modulators example

The amplitude scaling factors for the calculation of the FM spectrum of a single carrier with two parallel modulators scheme result from the multiplication of the respective Bessel functions: $B_n(i_1) \times B_m(i_2)$. In this case, the frequencies of the spectrum is calculated as follows:

$$f_c - (n \times f_{m1}) + (m \times f_{m2})$$
$$f_c - (n \times f_{m1}) - (m \times f_{m2})$$
$$f_c + (n \times f_{m1}) + (m \times f_{m2})$$
$$f_c + (n \times f_{m1}) - (m \times f_{m2})$$

Example:

$$i_1 = 1.5$$
$$i_2 = 1.0$$
$$f_c = 440\,\text{Hz}$$
$$f_{m1} = 100\,\text{Hz}$$
$$f_{m2} = 30\,\text{Hz}$$

$$B_n(1.5) \times B_m(1.0) \text{ for } 440 - (n \times 100) + (m \times 30)$$
$$440 - (n \times 100) - (m \times 30)$$
$$440 + (n \times 100) + (m \times 30)$$
$$440 + (n \times 100) - (m \times 30)$$

where $n = \{0, 1, 2\}$ and $m = \{0, 1, 2\}$.

The frequencies and amplitude scaling factors are calculated as follows (Figure A1.1):

(a) $B_0(1.5) \times B_0(1.0) = 0.512 \times 0.765 = 0.40$

$440 - (0 \times 100) + (0 \times 30) = 440\,\text{Hz}$

$440 - (0 \times 100) - (0 \times 30) = 440\,\text{Hz}$

$440 + (0 \times 100) + (0 \times 30) = 440\,\text{Hz}$

$440 + (0 \times 100) - (0 \times 30) = 440\,\text{Hz}$

(b) $B_0(1.5) \times B_1(1.0) = 0.512 \times 0.440 = 0.22$

$440 - (0 \times 100) + (1 \times 30) = 470\,\text{Hz}$

$440 - (0 \times 100) - (1 \times 30) = 410\,\text{Hz}$

$440 + (0 \times 100) + (1 \times 30) = 470\,\text{Hz}$

$440 + (0 \times 100) - (1 \times 30) = 410\,\text{Hz}$

(c) $B_0(1.5) \times B_2(1.0) = 0.512 \times 0.115 = 0.06$

$440 - (0 \times 100) + (2 \times 30) = 500\,\text{Hz}$

$440 - (0 \times 100) - (2 \times 30) = 380\,\text{Hz}$

$440 + (0 \times 100) + (2 \times 30) = 500\,\text{Hz}$

$440 + (0 \times 100) - (2 \times 30) = 380\,\text{Hz}$

(d) $B_1(1.5) \times B_0(1.0) = 0.558 \times 0.765 = 0.42$

$440 - (1 \times 100) + (0 \times 30) = 340\,\text{Hz}$

$440 - (1 \times 100) - (0 \times 30) = 340\,\text{Hz}$

$440 + (1 \times 100) + (0 \times 30) = 540\,\text{Hz}$

$440 + (1 \times 100) - (0 \times 30) = 540\,\text{Hz}$

(e) $B_1(1.5) \times B_1(1.0) = 0.558 \times 0.440 = 0.24$

$440 - (1 \times 100) + (1 \times 30) = 370\,\text{Hz}$

$440 - (1 \times 100) - (1 \times 30) = 310\,\text{Hz}$

$440 + (1 \times 100) + (1 \times 30) = 570\,\text{Hz}$

$440 + (1 \times 100) - (1 \times 30) = 510\,\text{Hz}$

(f) $B_1(1.5) \times B_2(1.0) = 0.558 \times 0.115 = 0.06$

$440 - (1 \times 100) + (2 \times 30) = 400\,\text{Hz}$

$440 - (1 \times 100) - (2 \times 30) = 280\,\text{Hz}$

$440 + (1 \times 100) + (2 \times 30) = 600\,\text{Hz}$

$440 + (1 \times 100) - (2 \times 30) = 480\,\text{Hz}$

(g) $B_2(1.5) \times B_0(1.0) = 0.232 \times 0.765 = 0.18$

$440 - (2 \times 100) + (0 \times 30) = 240\,\text{Hz}$

$440 - (2 \times 100) - (0 \times 30) = 240\,\text{Hz}$

$440 + (2 \times 100) + (0 \times 30) = 640\,\text{Hz}$

$440 + (2 \times 100) - (0 \times 30) = 640\,\text{Hz}$

(h) $B_2(1.5) \times B_1(1.0) = 0.232 \times 0.440 = 0.10$

$440 - (2 \times 100) + (1 \times 30) = 270\,\text{Hz}$

$440 - (2 \times 100) - (1 \times 30) = 210\,\text{Hz}$

$440 + (2 \times 100) + (1 \times 30) = 670\,\text{Hz}$

$440 + (2 \times 100) - (1 \times 30) = 610\,\text{Hz}$

(i) $B_2(1.5) \times B_2(1.0) = 0.232 \times 0.115 = 0.02$

$440 - (2 \times 100) + (2 \times 30) = 300\,\text{Hz}$

$440 - (2 \times 100) - (2 \times 30) = 180\,\text{Hz}$

$440 + (2 \times 100) + (2 \times 30) = 700\,\text{Hz}$

$440 + (2 \times 100) - (2 \times 30) = 580\,\text{Hz}$

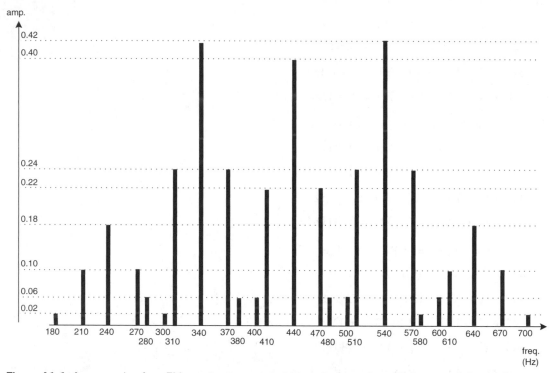

Figure A1.1 An example of an FM spectrum generated by a single carrier with two parallel modulators

Waveshaping

To a large extent the best transfer functions for waveshaping synthesis are described using polynomials. The amplitude of the signal input to the waveshaper is represented by the variable x and the output is denoted by $F(x)$, where F is the function and d are amplitude coefficients:

$$F(x) = d_0 + d_1x + d_2x^2 + \ldots + d_nx^n$$

Polynomials are useful because they can predict the highest partial that will be generated by the waveshaper. If the input is a sinusoid and the transfer function a polynomial of order n, then the waveshaper produces harmonics up to the nth harmonic. In this case, the

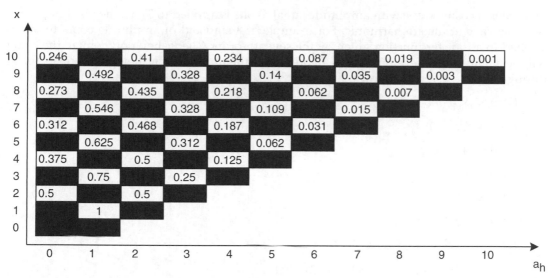

Figure A1.2 The amplitude of the harmonics produced by a term in the polynomial when the amplitude of the input sinusoid is equal to one

amplitudes of the harmonics can be calculated according the the table in Figure A1.2. This table lists the amplitudes of the harmonics produced by a term in the polynomial when the amplitude of the input sinusoid is 1. It also indicates that a term does not produce more harmonics than its exponent. For example, the transfer function $F(x) = x + x^2 + x^3 + x^4$ will produce the first, second, third and fourth harmonics. The amplitudes of the harmonics are calculated as the sum of the contributions of each term:

$a_{h0} = 0.5 + 0.375 = 0.875$

$a_{h1} = 1 + 0.75 = 1.75$

$a_{h2} = 0.5 + 0.5 = 1$

$a_{h3} = 0.25$

$a_{h4} = 0.125$

Note that the even power of x produces even harmonics, whilst the odd power produces odd ones. This affords musicians independent control of the odd and even harmonics of the sound.

When the waveform applied to the waveshaper is not a sinusoid, the resulting spectrum is more difficult to predict. The calculation of output spectra for non-sinusoidal input signal is beyond the scope of this book.

Chebyshev polynomials

Chebyshev polynomials are represented as follows: $T_k(x)$ where k represents the order of the polynomial and x represents a sinusoid. Chebyshev polynomials have the useful property that when a cosine wave with amplitude equal to one is applied to $T_k(x)$, the resulting signal is a sinewave at the kth harmonic. For example, if a sinusoid of amplitude equal to one is applied to a transfer function given by the seventh-order Chebyshev polynomial, the result will be a sinusoid at seven times the frequency of the input. Chebyshev polynomials for T_1 through T_{10} are given as follows:

$$T_1(x) = x$$

$$T_2(x) = 2x^2 - 1$$

$$T_3(x) = 4x^3 - 3x$$

$$T_4(x) = 8x^4 - 8x^2 + 1$$

$$T_5(x) = 16x^5 - 20x^3 + 5x$$

$$T_6(x) = 32x^6 - 48x^4 + 18x^2 - 1$$

$$T_7(x) = 64x^7 - 112x^5 + 56x^3 - 7x$$

$$T_8(x) = 128x^8 - 256x^6 + 160x^4 - 32x^2 + 1$$

$$T_9(x) = 256x^9 - 576x^7 + 432x^5 - 120x^3 + 9x$$

$$T_{10}(x) = 512x^{10} - 1280x^8 + 1120x^6 - 400x^4 + 50x^2 - 1$$

Because each separate polynomial produces a particular harmonic of the input signal, a certain spectrum composed of various harmonics can be obtained by adding a weighted combination of Chebyshev polynomials, one for each desired harmonic. For example, the transfer function to produce a spectrum containing the first, second, third and fifth harmonics can be determined as follows:

$$F(x) = w_1 T_1(x) + w_2 T_2(x) + w_3 T_3(x) + w_5 T_5(x)$$

The weighting values (w_1, w_2, w_3 and w_5) correspond to the relative amplitudes of the respective harmonic components.

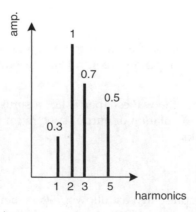

Figure A1.3 An example of a spectrum

As an example, consider the calculation of the tranfer function to produce the spectrum shown in Figure A1.3:

$$F(x) = 0.3T_1(x) + 1T_2(x) + 0.7T_3(x) + 0.5T_5(x)$$

$$F(x) = 0.3(x) + 1(2x^2 - 1) + 0.7(4x^3 - 3x) + 0.5(16x^5 - 20x^3 + 5x)$$

$$F(x) = 0.3x + 2x^2 - 1 + 2.8x^3 - 2.1x + 8x^5 - 10x^3 + 2.5x$$

$$F(x) = 8x^5 - 7.2x^3 + 2x^2 - 0.7x - 1$$

APPENDIX 2 Formant values

Vowel /a/ as in the word 'car' (in English):
Male:

F1 = 622.25 Hz	BW1 = 60 Hz	A1 = 0 dB
F2 = 1046.5 Hz	BW2 = 70 Hz	A2 = −7 dB
F3 = 2489 Hz	BW3 = 110 Hz	A2 = −9 dB

Female:

F1 = 783.99 Hz	BW1 = 80 Hz	A1 = 0 dB
F2 = 1174.7 Hz	BW2 = 90 Hz	A2 = −4 dB
F3 = 2793.8 Hz	BW3 = 120 Hz	A3 = −20 dB

Vowel /e/ as in the word 'bed' (in English):
Male:

F1 = 392 Hz	BW1 = 40 Hz	A1 = 0 dB
F2 = 1661.2 Hz	BW2 = 80 Hz	A2 = −12 dB
F3 = 2489 Hz	BW3 = 100 Hz	A2 = −9 dB

Female:

F1 = 392 Hz	BW1 = 60 Hz	A1 = 0 dB
F2 = 1568 Hz	BW2 = 80 Hz	A2 = −24 dB
F3 = 2793.8 Hz	BW3 = 120 Hz	A3 = −30 dB

Vowel /i/ as in the word 'it' (in English):
Male:

F1 = 261.63 Hz	BW1 = 60 Hz	A1 = 0 dB
F2 = 1760 Hz	BW2 = 90 Hz	A2 = −30 dB
F3 = 2489 Hz	BW3 = 100 Hz	A2 = −16 dB

Female:

F1 = 349.23 Hz	BW1 = 50 Hz	A1 = 0 dB
F2 = 1661.2 Hz	BW2 = 100 Hz	A2 = −20 dB
F3 = 2793.8 Hz	BW3 = 120 Hz	A3 = −30 dB

Vowel /o/ as in the word 'hot' (in English):
Male:

F1 = 392 Hz	BW1 = 40 Hz	A1 = 0 dB
F2 = 783.99 Hz	BW2 = 80 Hz	A2 = −11 dB
F3 = 2489 Hz	BW3 = 100 Hz	A2 = −21 dB

Female:

F1 = 440 Hz	BW1 = 70 Hz	A1 = 0 dB
F2 = 783.99 Hz	BW2 = 80 Hz	A2 = −9 dB
F3 = 2793.8 Hz	BW3 = 100 Hz	A3 = −16 dB

Vowel /u/ as in the word 'rule' (in English):
Male:

F1 = 349.23 Hz	BW1 = 40 Hz	A1 = 0 dB
F2 = 587.33 Hz	BW2 = 80 Hz	A2 = −20 dB
F3 = 2489 Hz	BW3 = 100 Hz	A2 = −32 dB

Female:

F1 = 329.63 Hz	BW1 = 50 Hz	A1 = 0 dB
F2 = 739.99 Hz	BW2 = 60 Hz	A2 = −12 dB
F3 = 2793.8 Hz	BW3 = 170 Hz	A3 = −30 dB

APPENDIX 3 Artist's Inductive Machine Learning Algorithm

To construct a decision tree DT from a training set TSet do:

1. If **TSet** is empty then **DT** is a single-node tree labelled **null**
2. Else
   2.1. If all the examples in **TSet** belong to the same sound class **SOUND_CLASS**
   2.2. Then **DT** is a single-node tree labelled **SOUND_CLASS**
   2.3. Else select the most informative attribute **MIA**
      2.3.1. If there is no **MIA** to choose
      2.3.2. Then **DT** is a single-node tree with the list of the conflicting examples
      2.3.3. Else
         2.3.3.1. From the **MIA** obtain its attribute values **atv(1)**, **atv(2)**, . . ., **atv(n)**
         2.3.3.2. Partition **TSet** into **TSet(1)**, **TSet(2)**, . . ., **TSet(n)**, according to the attribute values **atv** of **MIA**
         2.3.3.3. Construct recursively sub-decision-trees **ST(1)**, **ST(2)**, . . ., **ST(n)** for **TSet(1)**, **TSet(2)**, . . ., **TSet(n)**
         2.3.3.4. The result is the tree **DT** whose root **MIA** and whose sub-decision-trees are **ST(1)**, **ST(2)**, . . ., **atv(2)**, . . ., **atv(n)**

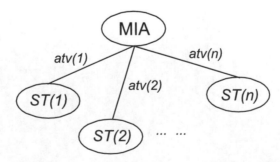

Figure A3.1 Each time a new MIA is selected the algorithm constructs recursively sub-decision-trees ST for each attribute of the MIA

Each time a new MIA (Most Informative Attribute) is selected, only those attributes that have not yet been selected in previous recursion (that is, used in the upper parts of the tree) are considered (Figure A3.1). When the available attributes are insufficient to distinguish between classes of training sound examples (that is, training sound examples that belong to different classes may have exactly the same attributes) then we say that these are conflicting training examples. If the algorithm cannot find a new MIA, then it records a list of conflicting training examples together with the number of occurrences in TSet of each element of the conflicting list. This information is used as a weight if a selection among them is eventually required.

References

Adrien, J.-M. (1991). The missing link: modal synthesis. In De Poli, G. *et al.* (eds), *Representation of Musical Signals*, Cambridge, MA: The MIT Press.

Antunes, J. (1995). Síntese sonora com harmônicos escorregadios, *II Simpósio Brasileiro de Computação e Música*, Canela, Brazil, pp. 32–35.

Arcela, A. (1994). A linguagem SOM-A para síntese aditiva, *I Simpósio Brasileiro de Computação e Música*, Caxambú, Brazil, pp. 33–43.

Arfib, D. (1979). Digital synthesis of complex spectra by means of multiplication of non-linear distorted sound waves, *Journal of the Audio Engineering Society*, **27**, 757–768.

Bailey, N. *et al.* (1990). Concurrent Csound: parallel execution for high speed direct synthesis, *Proceedings of the International Computer Music Conference*, Glasgow.

Bayle, F. (1993). *Musique Acousmatique*, Paris: INA/GRM – Edition Buchet/Chastel.

Berg, P. (1979). PILE – A language for sound synthesis, *Computer Music Journal*, **3**, 1, 30–41.

Bilotta, E., Miranda, E. R., Pantano, P. and Todd, P. M. (Eds.) (2001). *ECAL/ALMMA 2001: Workshop on Artificial Models for Musical Applications*, Cosenza: Editoriale Bios.

Bismark, G. von (1974). Timbre of steady sounds: a factorial investigation of its verbal attributes, *Acústica*, **30**, 146–158.

Boersma, P. (1991). Synthesis of speech sounds from a multi-mass model of the lungs, vocal tract, and glottis, *Proceedings of the Institute of Phonetic Sciences*, **15**, University of Amsterdam.

Boersma, P. and Weenink, D. (1996). *Praat, a system for doing phonetics by computer*. Technical Report 132, Institute of Phonetic Sciences of the University of Amsterdam.

Boulanger, R. (ed.) (2000). *The Csound Book*, Cambridge, MA: The MIT Press.

Bratko, I. (1990). *Prolog Programming for Artificial Intelligence*, Wokingham: Addison-Wesley.

de Bruin, G. and van Walstijn, M. (1995). Physical models of wind instruments: a generalised excitation coupled with a modular tube simulation platform, *Journal of New Music Research*, **24**, 148–163.

Cadoz, C. and Florens, J. L. (1990). Modular modelisation and simulation of the instrument, *Proceedings of the 1990 International Computer Music Conference*, ICMA.

Cadoz, C. and Florens, J. L. (1991). The physical model, modelisation and simulation systems of the instrumental universe. In De Poli, G. *et al.* (eds), *Representation of Musical Signals*, Cambridge, MA: The MIT Press.

Cadoz, C., Luciani, A. and Florenz, J. L. (1984). Responsive input devices and sound synthesis by instrumental mechanisms, *Computer Music Journal*, **8**, 3, 60–73.

Carbonell, J. G. (1990). *Machine Learning: Paradigms and Methods*, Cambridge, MA: The MIT Press – Elsevier.

Chandra, A. (1994). The linear change of waveform segments causing non-linear changes of timbral presence, *Contemporary Music Review: Timbre Composition in Electroacoustic Music*, **10**, 2, 157–169.

Chareyron, J. (1990). Digital synthesis of self-modifying waveforms by means of linear automata, *Computer Music Journal*, **14**, 4, 25–40.

Chowning, J. (1973). The synthesis of complex audio spectra by means of frequency modulation, *Journal of the Audio Engineering Society*, **21**, 7, 526–534.

Chowning, J. and Bristow, D. (1986). *FM Theory and Applications*, Tokyo: Yamaha Music Foundation.

Cook, P. (1992). SPASM: a real-time vocal tract physical model controller; and singer, the companion software synthesis system, *Computer Music Journal*, **17**, 1, 30–44.

Darwin, C. (1859). *On the Origins of species by means of natural selection or the representation of favoured races in the struggle for life*, London: John Murray.

Dawkins, R. (1986). *The Blind Watchmaker*, London: Penguin Books.

Deutch, D. (ed.) (1982). *The Psychology of Music*, Orlando, FL: Academic Press.

Dietterich, T. and Michalski, R. (1981). Inductive learning of structural descriptions, *Artificial Intelligence*, **16**.

Dodge, C. and Jerse, T. A. (1985). *Computer Music – synthesis, composition and performance*, New York, NY: Schirmer Books.

Ehresman, D. and Wessel, D. (1978). *Perception of Timbral Analogies*, Rapports Ircam, Paris: Centre Georges Pompidou.

Gabor, D. (1947). Acoustical quanta and the theory of hearing, *Nature*, **159**, 591–594.

Garcia, R. (2001). Growing sound synthesisers using evolutionary methods. In Bilotta, E., Miranda, E. R., Pantano, P. and Todd, P. M. (eds), *ALMMA 2001 – Workshop on Artificial Life Models for Musical Applications*, Cosenza: Editorial Bios.

Gawthrop, P. J. (1995). Physical model-based control: a bond graphs approach, *Journal of the Franklin Institute*, **332B**, 3, 285–305.

Gawthrop, P. J. and Smith, L. P. S. (1996). *Metamodeling: Bond Graphs and Dynamic Systems*, Hemel Hempstead: Prentice Hall.

Helmholtz, H. L. F. (1885). *On the sensations of tone as a physiological basis for the theory of music*, London: Longmans, Green and Co.

Holland, J. H. (1975). *Adaptation in Natural and Artificial Systems*, Ann Arbor, MI: University of Michigan Press.

Holtzman, S. R. (1978). *Description of an automated digital sound synthesiser instrument*, DAI Research Paper No. 59, Department of Artificial Intelligence, University of Edinburgh.

Horner, A., Beauchamp, J. and Haken, L. (1993). Machine Tongues XVI: Genetic algorithms and their application to FM matching synthesis, *Computer Music Journal*, **17**, 4, 17–29.

Howard, D. M. and Angus, J. (1996). *Acoustics and Psychoacoustics*, Oxford: Focal Press.

Hutchins, B. (1973). Experimental electronic music devices employing Walsh functions, *Journal of the Audio Engineering Society*, **21**, 8, 640–645.

Hutchins, B. (1975). Application of real-time Hadamard transform network to sound synthesis, *Journal of the Audio Engineering Society*, **23**, 5, 558–562.

Johnson, C. G. (1999). Exploring the sound-space of synthesis algorithms using interactive genetic algorithms, *AISB'99 Symposium on Musical Creativity*, Edinburgh, pp. 20–27.

Kaegi, W. and Tempellars, S. (1978). VOSIM – a new sound synthesis system, *Journal of the Audio Engineering Society*, **26**, 6, 418–426.

Karplus, K. and Strong, A. (1983). Digital synthesis of plucked string and drum timbres, *Computer Music Journal*, **7**, 2, 43–55.

Klaat, D. (1980). Software for a cascade/parallel formant synthesiser, *The Journal of the Acoustical Society of America*, **63**, 3, 971–995.

Koza, J. (1992). *Genetic Programming: On the Programming of Computers by Means of Natural Selection*, Cambridge, MA: The MIT Press.

Koza, J. R., Bennett III, F. H., Andre, D. and Keane, M. (1999). The design of analogue circuits by means of genetic programming. In Bentley, P. (ed.), *Evolutionary Design by Computers*, San Francisco, CA: Morgan Kaufmann.

Ladefoged, P. (2001). *Vowels and Consonants – An introduction to the sounds of language*, Malden, MA: Blackwell Publishers.

Laver, J. (1994). *Principles of Phonetics*, Cambridge: Cambridge University Press.

Lu, H.-L. and Smith, J. O. (2000). Glottal source modelling for singing voice synthesis, *Proceedings of the 2000 International Computer Music Conference*, pp. 90–97.

Luger, G. F. and Stubblefield, W. A. (1989). *Artificial Intelligence and the Design of Expert Systems*, Redwood City, CA: The Benjamin/Cummings Publishing Company.

Maconie, R. (1976). *The Works of Stockhausen*, London: Marion Boyars.

Maconie, R. (1989). *Stockhausen on Music*, London: Marion Boyars.

Maeda, S. (1982). A digital simulation method of the vocal-tract system, *Speech Communication*, **1**, 19–48.

Manning, P. (1987). *Electronic and Computer Music*, Oxford: Oxford University Press.

Mathews, M. (1969). *The Technology of Computer Music*, Cambridge, MA: The MIT Press.

McAdams, S. and Deliege, I. (issue eds) (1985). *Contemporary Music Review: Music and the Cognitive Sciences*, **4**.

McGregor, A., Miranda, E. R. and Gawthrop, P. (1999). A bond graphs approach to physical modelling of musical instruments, *Proceedings of the VI Brazilian Symposium on Computer Music*, Sociedade Brasileira de Computação (SBC), pp. 17–26.

Miller, J. and van Loon, B. (eds) (1990). *Darwin for Beginners*, New York: Pantheon Books.

Miranda, E. R. (1993). Cellular automata music: an interdisciplinary project, *Interface* (now called *Journal of New Music Research*), **22**, 3–21.

Miranda, E. R. (1994). *Sound Design: An Artificial Intelligence Approach*, PhD thesis, Edinburgh: University of Edinburgh.

Miranda, E. R. (1995a). Chaosynth: Um sistema que utiliza um autômato celular para sintetizar partículas sônicas, *II Simpósio Brasileiro de Computação e Música*, Canela, Brazil, pp. 205–212.

Miranda, E. R. (1995b). An artificial intelligence approach to sound design, *Computer Music Journal*, **19**, 2, 59–75.

Miranda, E. R. (1997). Machine learning of sound attributes: a case study, *Leonardo Music Journal*, **7**, 47–54.

Miranda, E. R., Correa, J. and Wright, J. (2000). Categorising complex dynamic sounds, *Organised Sound*, **5**, 2, 95–102.

Miranda, E. R. (2002). Generating source-streams for extralinguistic utterances, *Journal of the Audio Engineering Society*, **50**, 3, 165–172.

Moore, F. R. (1990). *Elements of Computer Music*, Englewood Cliffs, NJ: Prentice Hall.

Norman, K. (issue ed.) (1996). *Contemporary Music Review: A Poetry of Reality: Composing with Recorded Sound*, **15**, 1–2.

Olson, H. F. (1952). *Musical Engineering*, New York: McGraw-Hill.

Paynter, H. M. (1961). *Analysis and Design of Engineering Systems*, Cambridge, MA: The MIT Press.

Quinlan, J. R. (1986). Induction of decision trees, *Machine Learning*, **1**, 1.

Risset, J.-C. (1992). Timbre et synthèse des sons. In Barrière, J.-B. (ed.), *Le timbre: métaphore pour la composition*, Paris: Ircam and Christian Bourgois Editeur.

Roads, C. (1988). Introduction to granular synthesis, *Computer Music Journal*, **12**, 2, 11–13.

Roads, C. (2001) Sound composition with pulsars, *Journal of the Audio Engineering Society*, **49**, 3, 134–147.

Rodet, X. (1984). Time-domain formant-wave-function synthesis, *Computer Music Journal*, **8**, 3, 9–14.

Rossing, T. (1990). *The Science of Sound*, Reading, MA: Addison-Wesley.

Rumsey, F. (1996). *The Audio Workstation Handbook*, Oxford: Focal Press.

Rumsey, F. (1994). *MIDI Systems and Control*, Oxford: Focal Press.

Rumsey, F. and Watkinson, J. (1995). *The Digital Interface Handbook*, 2nd edition, Oxford: Focal Press.

Russ, M. (1996). *Sound Synthesis and Sampling*, Oxford: Focal Press.

Schaeffer, P. (1966). *Traité des objets musicaux*, Paris: Editions du Seuil.

Schottstaedt, B. (1994). Machine Tongues XVII: CLM: Music V meets Common Lisp, *Computer Music Journal*, **18**, 2, 30–37.

Serra, M.-H. (1993). *Stochastic Composition and Stochastic Timbre: GENDY 3 by Iannis Xenakis*, Paris: Cemamu.

Serra, X. and Smith, J. (1990). Spectral modelling synthesis: a sound analysis/synthesis system based on a deterministic plus stochastic decomposition, *Computer Music Journal*, **14**, 4, 12–24.

Slawson, W. (1985). *Sound Color*, Berkeley, CA: University of California Press.

Smith, J. (1992). Physical modeling using digital waveguides, *Computer Music Journal*, **16**, 4, 74–91.

Sobel, M. G. (1997). *A Practical Guide to Linux*, Boston, MA: Addison-Wesley.

Sterling, T., Salmon, J., Becker, D. J. and Savarese, D. F. (eds) (1999). *How to Build a Beowulf: A Guide to the Implementation and Application of PC Clusters*, Cambridge, MA: The MIT Press.

Stockhausen, K. (1959). . . . how time passes . . ., *Die Reihe*, 5, English edition, pp. 59–82.

Terhardt, B. (1974). On the perception of periodic sound fluctuation, *Acústica*, **30**, 201–213.

Truax, B. (1988). Real-time granular synthesis with a digital signal processor, *Computer Music Journal*, **12**, 2, 14–26.

Vaggione, H. (1993). Timbre as syntax: a spectral modeling approach, *Contemporary Music Review: Timbre Composition in Electroacoustic Music*, **10**, 2, 73–83.

Vercoe, B. (1986). *The Csound Manual*, Cambridge, MA: MIT Media Lab.

Vince, A. J. and Morris, C. A. N. (1990). *Discrete Mathematics for Computing*, Chichester: Ellis Horwood Limited.

Wessel, D. (1979). *Timbre space as a musical control structure*, Rapports Ircam, Paris: Centre Georges Pompidou.

Wiener, N. (1964). Spatial–temporal continuity, quantum theory, and music. In Capeck, M. (ed.), *The Concepts of Space and Time*, London: Reidel.

Winston, P. (1985). Learning structural descriptions from examples. In Brachman, R. and Levesque, H. (eds), *Readings in Knowledge Representation*, San Francisco, CA: Morgan Kaufmann Publishers.

Wishart, T. (1985). *On Sonic Art*, York: Imagineering Press.

Wishart, T. (1994). *Audible Design*, York: Orpheus the Pantomime Ltd.

Wolfram, S. (1986). *Theory and Applications of Cellular Automata*, Singapore: World Scientific.

Xenakis, I. (1971). *Formalized Music*, Bloomington, IN: Indiana University Press.

CD-ROM instructions

ATTENTION:

ALL MATERIALS ON THIS CD-ROM ARE USED AT YOUR OWN RISK. FOCAL PRESS AND THE AUTHOR OF THIS BOOK CANNOT ACCEPT ANY LIABILITY FOR ANY SOFTWARE MALFUNCTION OR FOR ANY PROBLEM WHATSOEVER CAUSED BY ANY OF THE SYSTEMS ON THIS CD-ROM.

ANY COMPLAINTS OR ENQUIRIES RELATING TO ANY SYSTEM OR DOCUMENTATION ON THIS CD-ROM MUST BE ADDRESSED DIRECTLY TO THE SOFTWARE AUTHORS AND NOT TO FOCAL PRESS.

ALL TEXTS ACCOMPANYING SOFTWARE ON THIS CD-ROM ARE PROVIDED AS GIVEN BY THEIR RESPECTIVE AUTHORS.

NOTE THAT SOME OF THE PROGRAMS ON THIS CD-ROM ARE EXPERIMENTAL PROTOTYPES, KINDLY PROVIDED BY THEIR AUTHORS PRIOR TO PUBLIC OR COMMERCIAL RELEASE. ALTHOUGH THEY WORKED PERFECTLY ON THE MACHINES USED TO TEST THEM, THERE IS NO ABSOLUTE GUARANTEE THAT THEY WILL RUN ON ALL COMPUTER CONFIGURATIONS AND UPGRADED VERSIONS OF THE OPERATIONAL SYSTEMS.

In order to run or install a program on your computer, you should normally copy the relevant folder onto your hard disk and follow the usual procedures for running or installing software. Most packages provide a Readme file with installation instructions; use Word Pad (on PC) or Simple Text (on Mac) to read a Readme file. A few general tips for each program are given below. It is strongly advised that you read them before you go on to use the programs.

Some packages provide HTML documentation. In principle you should be able to read them with any browser. They have been kept as simple as possible, stripped of fancy frames, backgrounds and tricky links. The objective is to convey the information hassle-free. A few packages provide documentation in PDF format. In order to read these files you will need to install Adobe Acrobat Reader; this is freely available from Adobe's Web site: <http://www.adobe.com>

Note that folders, file names and Web addresses are written between < >.

The folder <various> contains miscellaneous materials other than software. Cool Edit 98 (PC partition) is not a synthesis program. Rather, it is a sound editor which can be very useful for handling your sound files.

The folder <genesis> does not contain software, but sound examples and figures.

PC software

1. pcmusic <pcmusic>

This is a fully working sound synthesis programming language package. It does not need installation as such; simply copy the whole folder onto your hard disk.

The <Readme.txt> file inside the <pcmusic\Pcm> folder gives sufficient information to allow the sample scores to be compiled. However, be aware that the compressed file pcm09.exe is not on the CD-ROM, as it has already been expanded and all the files it would have generated are on the CD-ROM instead.

The file <pcmusic.pdf> contains a comprehensive tutorial furnished with a complete reference manual specifically prepared for this book by composer Dr Robert Thompson.

2. Nyquist <nyquistpc>

This is a fully fledged freeware sound synthesis and composition programming language. Nyquist is a very powerful tool for composition as it combines the features of synthesis-only languages (such as Csound) with composition-oriented capabilities. Nyquist does not need installation as such; simply copy the whole <nyquistpc> folder onto your main hard disk.

You will find the executable <nyquist.exe> in the <runtime> folder. Run this program and you will see a text window with some indication that Nyquist has started and has loaded some files. At this point the Nyquist is ready to take in Nyquist or Lisp expressions such as the ones given in the examples in section 1.3 of the user manual: <nyqman.pdf>.

Once you are ready to do serious work with Nyquist, you will probably want to keep your projects and compositions separate from the <runtime> folder where Nyquist is presently located; <introduction.html> explains how to set this up.

There are additional tutorials and documentation, kindly provided by Pedro J. Morales, in the folder <nyquistpc\demos>. Here you will find a number of examples of how to explore the potential of Nyquist. Also, a number of Nyquist programs are available in the folder <nyquist\test>.

3. Som-A <soma>

This is a fully working program and it does not need installation; simply copy the whole folder onto your hard disk.

If you repeatedly receive error messages when trying to open/save files, the best procedure is to delete the <Soma.cfg> file and configure the paths and directories from scratch; refer to the <Readme.txt> file for more instructions.

4. Audio Architect <audiarch>

In order to install the program, double click on <Setup.exe> and follow the procedures. You should not need to copy the folder onto your hard disk; <Setup.exe> should run fine from the CD-ROM. Note, however, that the folder <Aanets> containing the examples mentioned in the book should be copied manually because this folder is not part of the installation software.

The Quick Start section in the on-line Help is sufficient to allow you to listen to all the examples. The tutorial in the folder <Tutorial> is a very good way to get started with the program; it is excellent at taking over from where the Beginner's Guide section in the on-line Help finishes.

This is an unregistered demo version of the program but you should be able to make (and save) your own instruments with very few restrictions.

5. Reaktor <native>

In order to install the demo version or Reaktor, double click on <Reaktor3Demo.exe> and follow the installation instructions.

Keep an eye on Native Instruments' Web site for updated versions of their systems.

6. NI-Spektral Delay <native>

The NI-SPEKTRAL DELAY runs on Macintosh or under Windows, either stand-alone with MME, Direct Sound, Sound Manager and ASIO or as a plug-in with a VST – or DirectX – compatible host program.

In order to install the demo version of this software, double click on <spektraldemo.exe> and follow the installation instructions.

Refer to Native Instruments' Web site for updated versions of their systems.

7. Reality <reality>

In order to install the demo version of Reality, double click on <RealityDemo.exe> and follow the installation instructions.

A comprehensive user manual containing an introduction to the system is given in <manual.pdf>.

8. CDP and Sound Shaper <cdpdemo>

This package does not need installation as such; simply copy the whole folder onto your hard disk. Note that the CDP programs are designed to run under DOS (command-line interface) rather than Windows (but, see Sound Shaper below). You should understand DOS in order to set up the paths, etc. as outlined in the documentation. In order to run CDP under DOS, double click on the icon <CDP CommandLine>.

Bear in mind that output files cannot have the same name as previously existing ones; previous files must be deleted before saving. Output files must have the correct extension. Failure to observe these two points will produce an error message: 'failed to open xxx for output'.

Note that sound files and spectral analysis files both use the <.wav> extension. It might be a good idea to prefix all sound files with 's' and all analysis files with 'a'; e.g. <smysound.wav> and <amystuff.wav>.

The documentation for the CDPSPEC program suggests typing 'spec blur' from help with the BLUR function. You should instead type 'cdpspec blur'; this applies for all CDPSPEC functions.

Refer to the <README.txt> file for more installation instructions.

9. Sound Shaper <cdpdemo\Sndshapr>

This is a demo version of a front end graphic interface for the CDP package. It does not need installation; simply copy the whole folder onto your hard disk. It runs straightaway on a stand-alone PC, without any special setting up required.

An excellent tutorial specifically prepared for this book by composer Dr Robert Thompson is provided in <tutorial.pdf>. Comprehensive documentation is also available in the folder <cdpdemo\Sndshapr\docs>. Also, refer to the <README.txt> file in the folder <cdpdemo> for more instructions.

10. Wigout and TrikTraks <wigtrik>

This is a fully working package. It does not need installation as such; simply copy the whole folder onto your hard disk.

Start by reading the <Readme.rtf> file and then use your Web browser to read the <Index.html> file which is in the index for the whole HTML documentation. Do not change the original configuration of the directories as this will change the links between the HTML documents.

11. SMS Tools <sms>

This is a fully working program and it does not need installation as such; simply copy the whole folder onto your hard disk.

The document <Article.dpf> (also available in postscript and MS-Word formats) contains an in-depth explanation of the theory behind the system. Also, there is an HTML manual in the folder <sms\smstools\Manual.htm>.

12. Chaosynth <chaosynth>

This a demo version of the author's own software for granular synthesis.

In order to install the demo you will need the Microsoft Installer, which is normally supplied with Windows ME, 2000 or XP. Is also comes with other Microsoft packages such as Office. If your PC does not have this installer you need to download it from Microsoft and install it first. Please refer to the <README.txt> file for more information.

13. crusherXLive <crusherx>

In order to install this software, double click on <cxlSetup240.exe> and follow the installation instructions.

The <readme.htm> file that is installed inside the folder <crusherXLive> is the best place to start in order to learn about the system.

14. Virtual Waves <virwaves>

In order to install this software, double click on <Setup.exe> and follow the installation instructions. Note that the folder <Synthvir>, containing the examples mentioned in the book should be copied manually to your hard disk, as this folder is not part of the installable package.

15. Praat <praat>

This is a fully working version of the superb software originally designed as a tool for research in phonetics. In order to install Praat, copy the file <praat4004_winsit.exe> onto your hard disk and double click on it. Follow the instructions; click OK and a file called <Praat> or <Praat.exe> will be installed in the directory of your choice – this is the Praat program.

The user manual, accessed via the Help menu, is very complete and extremely helpful.

The folder <praat\scripts> contains examples of the scripts, including those mentioned in the book. You should copy this folder onto your hard disk manually as it is not part of the installable material.

Macintosh software

16. Nyquist <nyquimac>

This is a fully fledged sound synthesis and composition programming language. Nyquist is a very powerful tool for composition as it combines the features of synthesis-only languages (such as Csound) with composition-oriented capabilities. Nyquist does not need installation as such; simply copy the whole <nyquimac> folder onto your hard disk.

The executable file is in the folder <nyquimac:Nyquist:runtime>. Please read both <README> files in the folders <nyquist> and <nyquimac:Nyquist>. The user manual is very comprehensive and you will probably refer to it extensively if you use Nyquist for your projects. The manual is provided in three formats: HTML, PDF and TXT. In the folder <nyquimac:Nyquist:demos> there are additional tutorials and documentation specially prepared for this book. Here you will find many didactic examples to explore the potential of Nyquist. Also, a number of Nyquist programs are available in the folder <nyquimac:Nyquist:test>.

17. Reaktor <native>

In order to install the demo version of Reaktor, copy the file <Reaktor 3 Demo> onto your hard disk and double click on it. Then follow the installation instructions.

Before installing the demo make sure that your computer fulfills the following system requirements: Mac G3 or G4, 300 MHz or faster, 128 MB RAM. You will need a free USB port to run the FULL VERSION; the demo version does not require USB.

The <Readme.html> that is installed inside the folder <Reaktor 3 DEMO:Readme & Docs> contains all the information you need to get started; use a Web browser to read this document.

18. NI-Spektral Delay <native>

The NI-SPEKTRAL DELAY runs on Macintosh or under Windows, either stand-alone with MME, Direct Sound, Sound Manager and ASIO or as a plug-in with a VST – or DirectX – compatible host program.

In order to install the demo version of NI-Spektral Delay, copy the file <NI Spektral Delay Demo Install> onto your hard disk and double click on it. Then follow the installation instructions.

19. Praat <praat>

This is a fully version of the superb software originally designed as a tool for research in phonetics. There is no installation as such; just copy the <Praat 4.0> folder onto your Mac. The executable program is <Praat>. Just double click on it and it runs.

The only other file in the Praat distribution (in Folder Praat 4.0) is the READ ME file that you should read before using the system.

The user manual, accessed via the Help menu, is very complete and extremely helpful.

The folder <praat:scripts> contains examples of the scripts, including those mentioned in the book.

20. Diphone <diphone>

In order to install the Diphone demo, copy the folder <diphone> (containing the four <DiphoneDemo 1.7.1.sea> files) onto your hard disk and double-click on the self-extracting application <DiphoneDemo 1.7.1.sea.#1>. Then follow the instructions provided in the accompanying documentation.

21. LASy <lasy>

This is a fully working program and it does not need installation as such; simply copy the whole folder onto your hard disk. The accompanying documentation provides all necessary information for using the system.

22. Chaosynth <chaosynth>

This a demo Mac version of the author's own software for granular synthesis.

In order to install the system on your Mac simply copy the <cdemomac.sit> folder onto your computer, double click on it and follow the instructions.

Alternatively, just copy the folder <chaosynth:Chaosynth> onto your hard disk.

The user manual is available via the program's Help menu: 'Chaosynth Guide'. For more information (machine compatibility, etc.) contact Nyr Sound technical support: <support@nyrsound.com>.

The folders <chaosynth:taxonomy> and <chaosynth:classic-sounds> contain sound examples but these files are not part of the installation program. Make sure you have OMS installed on your computer in order to handle MiDi.

23. Pulsar <pulsar>

Pulsar Generator runs on PowerMacintosh 8600/9600, G3 and G4 computers, under Mac OS 8.5.1, 8.6, 9.0, 9.1. The demonstration version has no built-in recording capability and it generates no more than one minute of sound at a time. Also, it stops operating after 30 minutes or 75025 'pulsar credits'.

Launch the program by double-clicking on the <PulsarGenerator 2001 demo> icon.

24. Vibra1000 <koblo>

In order to install Koblo's Vibra1000 synthesiser double click on <Vibra1000 Installer> and follow the installation instructions. This will install a program called Tokyo and Vibra1000. Tokyo is a front end for running Koblo's applications, such as Vibra 1000. This is the fully working version of Tokyo and Vibra 1000.

In order to run Vibra1000, first run Tokyo and then select 'Vibra1000' from the Synths menu.

Index

Focal Press

www.focalpress.com
Join Focal Press on-line
As a member you will enjoy the following benefits:

- ⟨ an email bulletin with **information on new books**
- ⟨ a regular **Focal Press Newsletter**:
    - featuring a selection of new titles
    - keeps you informed of **special offers, discounts and freebies**
    - alerts you to **Focal Press news and events** such as author signings and seminars
- ⟨ complete access to **free content** and reference material on the focalpress site, such as the focalXtra articles and commentary from our authors
- ⟨ a **Sneak Preview** of selected titles (sample chapters) *before* they publish
- ⟨ a chance to have your say on our **discussion boards** and **review books** for other Focal readers

Focal Club Members are invited to give us feedback on our products and services.
Email: worldmarketing@focalpress.com – we want to hear your views!

Membership is **FREE**. To join, visit our website and register. If you require any further information regarding the on-line club please contact:

Lucy Lomas-Walker
Email: l.lomas@elsevier.com
Tel: +44 (0) 1865 314438
Fax: +44 (0)1865 314572
Address: Focal Press, Linacre House,
Jordan Hill, Oxford, UK, OX2 8DP

Catalogue
For information on all Focal Press titles, our full catalogue is available online at www.focalpress.com and all titles can be purchased here via secure online ordering, or contact us for a free printed version:

USA
Email: christine.degon@bhusa.com
Tel: +1 781 904 2607 T

Europe and rest of world
Email: j.blackford@elsevier.com
el: +44 (0)1865 314220

Potential authors
If you have an idea for a book, please get in touch: